Safety and Health Division 2015

Core Programming Area at the 2015 AIChE Annual Meeting

Salt Lake City, Utah, USA
8-13 November 2015

ISBN: 978-1-5108-1872-9

Printed from e-media with permission by:

Curran Associates, Inc.
57 Morehouse Lane
Red Hook, NY 12571

Some format issues inherent in the e-media version may also appear in this print version.

Printed by Curran Associates, Inc. (2016)

For permission requests, please contact AIChE
at the address below.

AIChE
120 Wall Street, FL 23
New York, NY 10005-4020

Phone: (800) 242-4363
Fax: (203) 775-5177

www.aiche.org

Additional copies of this publication are available from:

Curran Associates, Inc.
57 Morehouse Lane
Red Hook, NY 12571 USA
Phone: 845-758-0400
Fax: 845-758-2634
Email: curran@proceedings.com
Web: www.proceedings.com

TABLE OF CONTENTS

435013 Integration of Process Safety from Introduction to Chemical Engineering to Reaction Engineering

Abstract:

Process safety is essential to undergraduate chemical engineering education to adequately prepare students to be conservative, yet innovative chemical engineers in practice. Even at schools where industrial cooperative education is integral to a college degree, it is important to emphasize chemical process safety throughout the curriculum to reinforce safety topics exposed on the job, or due to the variety of employers, introduce students to concepts that they may not have experienced. It is important that chemical engineering students are exposed to process safety starting their freshman year to prepare them for cooperative employment and have this carry through to the senior year. At Kettering University students were introduced to industrial hygiene, risk management, and literature review on plant disasters in a new freshman introduction to chemical engineering course in Fall 2014. In unit operations and separations laboratory courses, students prepare safety case studies. Safety is further reinforced in reaction engineering with case studies, reviews of MSDS, examples of consequences from run-away reactions, and student-inspired safety presentations. An elective course covers risk management plans and the associated issues when considering the design and start-up of chemical plants. Finally graduating seniors have the opportunity to take an independent study course on Process Safety where the deliverable is case studies for use in lower-level courses. The integration and impact of how safety is presented in the Kettering University chemical engineering curriculum will be presented.

433695 Safety Immersion Education Using Vanderbilt's Chemical Process Innovation Center

Abstract:

Process and product safety are of increasing importance in the design and operation of chemical engineering processes. In general, a number of chemical process and product failures have been the result of design deficiencies such as the improper selection of materials of construction for the intended end use, the improper processing and/or fabrication of these materials into the end product and the improper maintenance of the chemical process and/or product, in addition to other factors. Furthermore, failure analysis of several chemical plant accidents has also resulted in the identification of numerous process design deficiencies. Scientific methodologies and tools used for identifying the potential mode of failures for polymer products and chemical processes including, but not limited to, what-if analysis, failure mode and effects analysis (FMEA) and hazard and operability analysis (HAZOP). Proper instruction and application of these methodologies, in addition to the use of case studies, are essential elements of the undergraduate chemical engineering curriculum. Specifically, efforts have been underway over the past two years to incorporate safety methodologies and a safety culture into the chemical engineering undergraduate laboratory courses and senior design courses. Using our newly renovated Chemical Process Innovation Center has further expanded safety education for chemical engineering students. Experience and lessons learned in applying and reinforcing safety in these courses will be presented along with a description of the Chemical Process Innovation Center.

435980 Process Safety Education Framework for Multidisciplinary Engineering Programmes

Abstract:

The rise in multidisciplinary engineering education across universities in the developed countries is significantly redefining approaches to teaching and learning. Safety education is a multidisciplinary subject linked directly to most engineering and science related undergraduate courses. With increasing significance in engineering degree courses, it has been particularly identified with Chemical, Process, and Petroleum Engineering related disciplines owing to the nature of associated hazards. This study provides a generic framework for adequate inclusions of process safety elements into the multidisciplinary engineering and science curriculum. The process safety elements are to be developed based on proposed Cogex framework integrated with interdisicplinary learning case studies. Safety education framework like this becomes necessary for an early exposure to core safety concerns and build up of cognitive understanding of work place practices as they relate to more complex engineering processes.

432853 From "Process Control" to "Process Control and Safety"

Abstract:

In order to address the ABET curriculum requirement to include the hazards associated with chemical processes, we have chosen at UMBC to modify our currently established Chemical Process Control course to be a course on Process Control and Safety. I will outline the major changes that were made to the course to accommodate this change in content.

The focus on the controls component of the course has become more practical and focused on the physical design of control schemes for unit operations, removing a bulk of the mathematics formally done “by hand” and instead using computer software to simulate and solve controls problems. There is still a semester project that requires the simulation of implemented PID controls, tuning, and stability analysis, but more time is devoted to literacy of process flow diagrams, control and block diagrams, and piping and instrumentation.

With the time freed from removing some weeks of mathematical analysis, the course can focus on further elements related to physical design, including placement and sizing of relief devices, analysis of source models, and prevention of fires and explosions. The safety component of the course culminates in the creation of a Hazards and Operability study for a very small process, as well as a requirement to inspect one lab on campus to practice using a safety checklist and identifying hazards.

As part of our documentation for ABET accreditation, I will outline how awareness of process safety is assessed, as well as how trends in responses from student surveys of graduating seniors has changed in light of the changes to this traditional process control course.

425684 Incorporating Safety into the Undergraduate Laboratory and Capstone Design Course

Abstract:

Safety training is an integral part of laboratory experimentation and chemical processing, and is a requirement of ABET. The challenge is to make safety an interesting aspect within the laboratory and classroom environments and to engage students. In addition to reviewing the Bhopal disaster and safe laboratory practices, it is important to provide classroom workshop type discussions where students can think and interact about "worst case scenarios" and "how to prevent them from ever happening" for specific processes and also as a requirement for their chemical capstone design project processes. Discussions should include current event safety incidents and also differences between voluntary and involuntary risk, and planned, fugitive, and emergency emissions. Laboratory experimentation should always include completing a pre-startup checklist before any experiments are carried out. Further, the safety program in research laboratories should be a shared responsibility with different safety responsibilities done by different researchers. This presentation will provide some useful examples and practices for the classroom and laboratory that the author has found to be valuable.

Simulation Solutions, Inc.

Minds-On/Hands-On Training for Operators™

"Integrating Simulation of Chemical Engineering Operations into Process Safety Education"

Mr. Donald C. Glaser
President

Mr. Matthew Garvey
Project Engineer

Simulation Solutions, Inc.
179 Avenue at the Common, Suite 1
Shrewsbury, NJ 07702-4804
dgalser@simulation-solutions.com

Robert G. Bozic, Ph.D.

LTC U.S. Army(Retired)
Lecturer in Chemical Engineering Design and
Chemical Engineering Process Safety
Department of Chemical Engineering
Columbia University in the City of New York
robert.bozic@columbia.edu

COLUMBIA ENGINEERING
The Fu Foundation School of Engineering and Applied Science

Outline

- Background/Motivation
- Safety Recommendations for ABET
- CHEN E4501 Chemical Engineering Process Safety
- SAChE Integration
- Learning from History (Case Study & Design)
- Using Simulations to Teach Process Safety
- Acknowledgements and References
- Conclusion and Future Work

2

Background and Motivation

"SAChE Recommendations for ABET Safety Content in Chemical Engineering February 18, 2010"
"The Safety and Chemical Engineering Education (SAChE) Committee makes the following recommendations with regard to student understanding to meet ABET curriculum requirements.

1. The graduate must understand the importance of process safety and the resources and commitment required. This should include the important incidents that define process safety, and how these incidents affected the practice of chemical engineering.

2. The graduate must be able to characterize the hazards associated with chemicals and other agents. This must include toxic, flammable, and reactive hazards.

3. The graduate must understand and be able to apply concepts of inherently safer design.

4. The graduate must understand how to control and mitigate hazards to prevent accidents. This should include generally accepted management systems, plant procedures and designs to prevent accidents.

5. The graduate should be familiar with the major regulations that impact the safety of chemical plants.

6. The graduate should understand the consequences of chemical plant incidents due to acute and chronic chemical releases and exposures.

7. The graduate should be reasonably proficient with at least one hazard identification procedure.

8. The graduate should have an introduction to the process of hazard evaluation and risk assessment. Resource materials to assist in meeting these requirements can be found on the SAChE web."

3

(www.sache.org accessed 13 Apr 2015)

Background and Motivation

"ABET Curriculum Statement approved in October 2011"

"1. Curriculum
The curriculum must provide a thorough grounding in the basic sciences including chemistry, physics, and biology, with some content at an advanced level, as appropriate to the objectives of the program. The curriculum must include the engineering application of these basic sciences to the design, analysis, and control of chemical, physical, and/or biological processes, including the hazards associated with these processes."

4

(www.abet.org accessed 13 Apr 2015)

CHEN E4501 Chemical Engineering Process Safety Course Description

"Aimed at seniors and graduate students. Provides classroom experience on chemical engineering process safety as well as Safety in Chemical Engineering certification. Process safety and process control emphasized. Application of basic chemical engineering concepts to chemical reactivity hazards, industrial hygiene, risk assessment, inherently safer design, hazard operability analysis, and engineering ethics. Application of safety to full spectrum of chemical engineering operations."

((http://www.columbia.edu/cu/bulletin/uwb accessed 27 Sep 2015)

CHEN E4501 Chemical Engineering Process Safety Course Significance

"Chemical Engineering Process Safety will be used anywhere in the world across the complete range of chemical engineering operations. Process Safety is current recommended ABET curriculum content by the Safety in Chemical Engineering Safety Committee. Academia and industry contributed to the content of applying chemical engineering concepts to chemical reactivity hazards, industrial hygiene, risk assessment, inherently safer design, and hazard operability analysis. With the ultimate aim of safer operations in the chemical engineering profession, the applications of the Chemical Engineering Process Safety will make the world safer."

5

(Fall 2015 Course Proposal)

CHEN E4501 Chemical Engineering Process Safety Course Objectives

1. Learn about the importance of process safety and the resources and the commitment required. This should include the important incidents that define process safety, and how these incidents affected the practice of chemical engineering.

2. Learn how to characterize the hazards associated with chemicals and other agents to include toxic, flammable, and reactive hazards.

3. Learn how to apply concepts of inherently safer design.

4. Learn how to control and mitigate hazards to prevent accidents.

5. Become familiar with the major regulations that impact the safety of chemical plants.

6. Learn about the consequences of chemical plant incidents due to acute and chronic chemical releases and exposures.

7. Become familiar with the process of hazard identification and evaluation, and risk assessment

8. Become familiar with engineering ethics.

6

(Spring 2015 CHEN 4501 Syllabus)

Course Design and SAChE Integration

Week 1: Process Safety and Course overview
Intro to Process Safety (Chp 1) and Intro to SAChE (www.sache.org)
Week 2: Chemical Reactivity Hazards
Chemical Reactivity (Chp 8) and Hazard Identification (Chp 11)
(SAChE Chemical Reactivity Hazards Completion certificate)
Week 3: Industrial Chemical Hygiene (Chp 3) and Toxicology (Chp 2)
(SAChE Chemical Process Safety in the Chemical Process Industries (Crowl) completion certificate)
Week 4: Risk Assessment (Chp 12)
(SAChE Risk Assessment completion certificate)
Week 5: Prevention of Fires and Explosions (Chp 6 and Chp 7)
(SAChE Dust Explosion Control (Louvar) completion certificate)
Week 6: Chemical Reactor Safety
(SAChE Runaway Reactions (Theis) completion certificate)
Week 7: Safety Procedure and Designs (Chp 13)
(SAChE Process Safety 101 (Bernardi) completion certificate)
Week 8: Inherently Safer Design
Intro to Reliefs (Chp 9), Relief Sizing (Chp 10)

(www.sache.org 27 Sep 2015)

Course Design and SAChE Integration

Week 9: Inherently Safer Design (Continued) and Engineering Ethics
Intro to Reliefs (Chp 9), Relief Sizing (Chp 10), Order of Engineer.
(SAChE Inherently Safer Design completion certificate)
Week 10: Intro to HAZOPs and Layer or Protection Analysis
(SAChE Process Safety Lessons Taught from Experience (Wiley))
Week 11: Guest lecture: **Simulation Solutions, Inc.**
Week 12: HAZOPS Project Progress Review
Week 13: HAZOPS Project Progress Review
Week 14: Presentation of Accident Investigation case studies
Week 15: Study Days
Week 16: Final HAZOPS Project Due

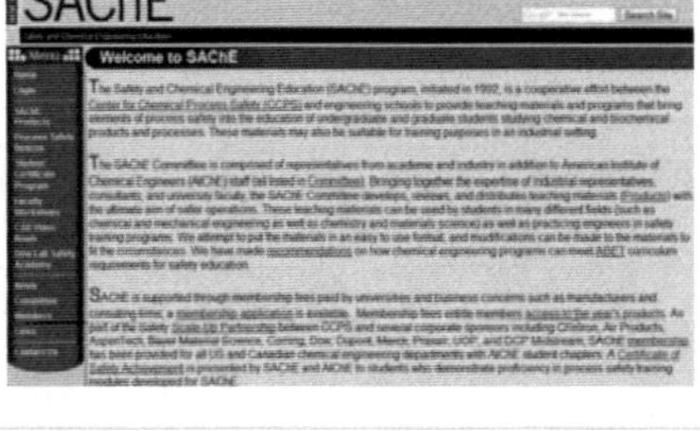

(www.sache.org 27 Sep 2015)

SAChE Integration Performance Review

	Bachelors		Masters		
Test	Avg	Stdev	Avg	Stdev	
1	0.89	0.078	0.93	0.054	SACE CP Dust Explosion Control - Final Exam
2	0.91	0.053	0.92	0.057	SACE CP Laboratory Safety - Final Exam
3	0.92	0.077	0.89	0.064	SAChE CP Chemical Reactivity Hazards - Quiz
4	0.89	0.067	0.91	0.061	SAChE CP Inherently Safer Design - Quiz
5	0.91	0.037	0.91	0.065	SAChE CP Process Safety 101 - Quiz
6	0.91	0.061	0.92	0.068	SAChE CP Process Safety Lessons Taught from Experience - Quiz
7	0.91	0.058	0.91	0.064	SAChE CP Risk Assessment - Quiz
8	0.89	0.054	0.92	0.056	SAChE CP Runaway Reactions - Quiz
9	0.94	0.067	0.93	0.060	SAChE CP Safety in the Chemical Process Industries - Quiz
10	0.88	0.072	0.93	0.066	SAChE Nitrogen Certificate - Final Exam
Overall	0.90	0.062	0.92	0.061	

Completion of the certification was used to calculate student grades. AIChE provided the raw data used to calculate the averages above.

9

(CHEN E4501 Course Assessment 27 Sep 2015)

Learning from History
Case Study and Design

Case study presentations
Spring 2015.

10

Learning from History
Case Study Format

Oral Communication Skills- (Briefing During Class)

Written Communication Skills- (Report turned in to the instructor)

Block Flow and Process Flow Diagrams

Case History (the accident, the event that produces injury/death/damage, and the hazard, the chemical or physical condition that can cause damage. Identify and evaluate the hazards involved. Determine what controls were used, and what should or could have been used. Define the scenario, a description of the events that result in an accident; the indecent, the loss of cantonment of material/energy, the consequence, result of accident, and the likelihood, expected probability of the event. Remember to address, risk- as defined, Risk = f(frequency, consequence) of the accident and determine what aspects of inherently safer design, moderate, substitute, minimize, and simplify, could have been used in order avoid such an accident)

Lessons Learned

Documentation/References

11

(Crowl, Daniel A. and Louvar, Joseph, 2011, Chp1)

Learning from History
Design

Oral Communication Skills - (In Class Progress Review)

Written Communication Skills- (Report turned in to the instructor)

Block Flow and Process Flow and Process Control Block Diagram Diagrams

Design Results (Outline what you recommend as a replacement and what your major conclusions are from your design and safety analysis. Design Results include summary of ASPEN simulation results, your safety analysis results from the Process Hazard Analysis Risk Assessment, Event Tree Analysis, Layer Of Protection Analysis, HAZOPS Analysis. Make sure you address aspects of Inherently Safer Design: Moderate, Substitute, Minimize, and Simplify. In this project, use Aspen Plus 8.6 for simulating your material and energy balances. As appropriate, make use of good engineering assumptions in order to complete this design project. Remember to assume, check, and validate assumptions. Keep a record of your assumptions and include them in your presentation.)

Discussion (Discussion of major decisions for your design results and your safety analysis)

Documentation/References

12

(Crowl, Daniel A. and Louvar, Joseph, 2011, Chp1)

9 Significant Disasters (Case Studies)

Learning from History

Louisville, Kentucky Mar 2011 (explosion)

Anacortes, Washington Apr 2010 (explosion)

Flixborough, England, June 1974 (pipe rupture)

Port Wentworth, Georgia Feb 2008 (dust explosion)

Jacksonville, Florida, Dec 2007 (explosion)

Texas City, Texas May, 2005 (explosion)

Pasadena, Texas, Oct 1989 (explosion)

Bhopal, India, Dec 1984 (toxic release)

Seveso, Italy, Jun 1976 (runaway reactor, toxic release)

(Crowl, Daniel A. and Louvar, Joseph, 2011, Chp1)

13

Chemical Engineering Safety Training Results

COLUMBIA UNIVERSITY

Donald C. Glaser & Matthew Garvey
Simulation Solutions, Inc.

April 7th & 9th 2015

14

Objectives

- Learn more about the importance of process safety and the resources & commitment required.
- Learn how to control and mitigate hazards to prevent accidents.
- Learn about operator training through the use of simulators.
- Review process control and chemical engineering design concepts.

15

Simulation Solutions, Inc. Outline

- Overview of Operator Training
- Review of Operator Safety & INSTO™ Methodology
- Introduction to P&ID / DCS / Outside Operator
- SSI "What-If" Exercises
- SSI Troubleshooting Exercises
- Q&A for Students

16

Simulators provide life-like Training for New and Experienced Operators

Laboratory Experience

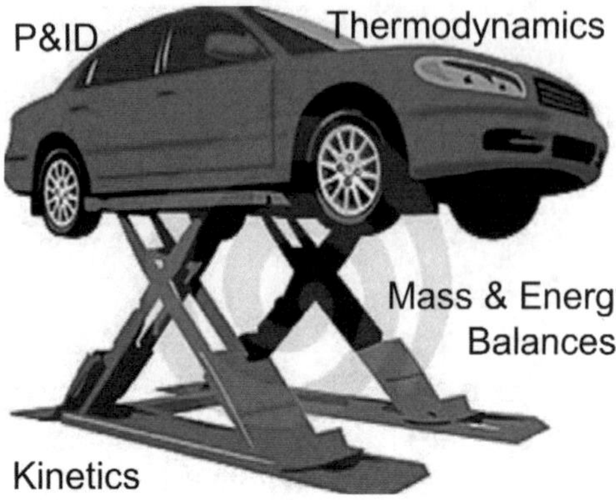

Real World Experience

17

Minds-On/Hands-On Troubleshooting Process

"Minds-On"

- Identify the Problem
- Troubleshooting Process
- Propose a Cure

"Hands-On"

- Taking Action
  - Safe Shutdown Or Corrective Action

18

Grafting an *Operations Mindset*™ to New Operators

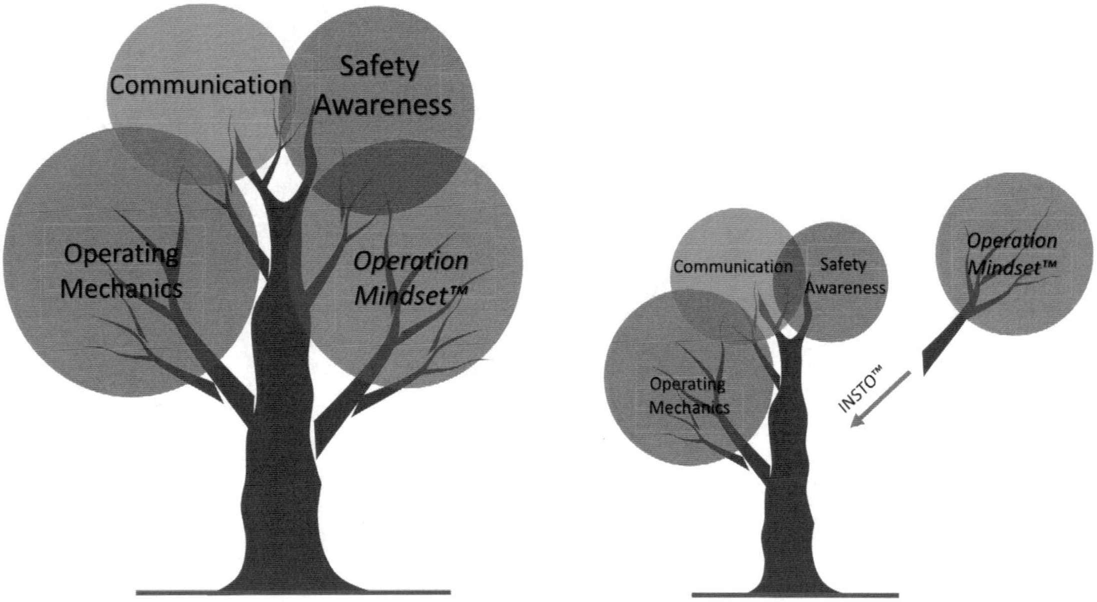

"Old School" Operator
Numerous Opportunities for Experiential Learning

"New School" Operator
Vastly Different Circumstances

19

Simulator Demonstrations

Day 1 – Heat Exchanger

Day 2 – Distillation

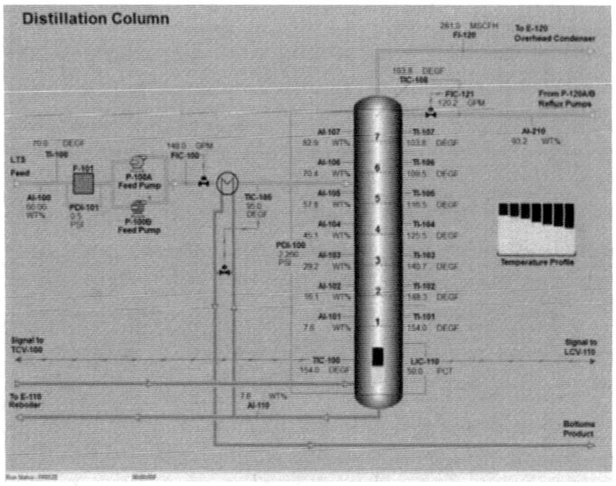

20

21

22

23

24

25

Student Survey Comments

- "Enjoyed seeing the practical implication of understanding these controls and how they are relevant to the industry operations."
- "I enjoyed the what-if analysis that helped me see the consequence of a single process change to the entire unit."
- "It was a great experience doing the exercises to understand what changes when you manipulate different variables pertaining to a heat exchanger or distillation column."

26

Student Survey Results (Graded on Scale 1-5)

Q1	As a result of CHEN E4501 lessons 21 and 22 my understanding of process control and chemical engineering design concepts has improved.
Q2	As a result of lessons 21 and 22 my understanding of operator training through the use of simulators has improved.
Q3	As a result of CHEN E4501 lessons 21 and 22 my ability to understand how to control and mitigate hazards to prevent accidents has improved. (course objective 4)
Q4	As a result of CHEN E4501 lessons 21 and 22 my ability to understand the importance of process safety and the resources and commitment required has improved. (course objective 1)

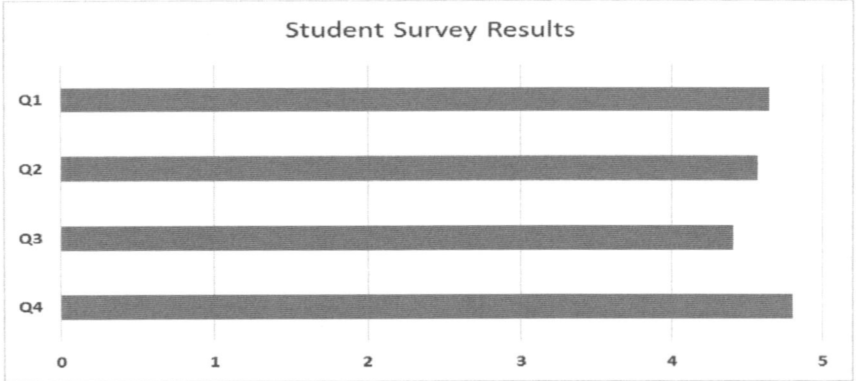

rank from 1-5, where 5 indicates strong agreement and 1 indicates strong disagreement.

27

Student Survey Results

For each statement below, rank from 1-5, where 5 indicates strong agreement and 1 indicates strong disagreement.	CHEN E4501 Lesson 21 and 22	Standard Deviation
As a result of CHEN E4501 lessons 21 and 22 my ability to understand the importance of process safety and the resources and commitment required has improved. (course objective 1)	4.80	0.50
As a result of CHEN E4501 lessons 21 and 22 my ability to understand how to control and mitigate hazards to prevent accidents has improved. (course objective 4)	4.40	0.58
As a result of lessons 21 and 22 my understanding of operator training through the use of simulators has improved.	4.56	0.65
As a result of CHEN E4501 lessons 21 and 22 my understanding of process control and chemical engineering design concepts has improved.	4.64	0.57

28

Comments from current Chemical Engineering Bachelors and Masters Students:

- "Quite useful for process control."
- "I think it is greater to know more about the industry and see how things work. It's not complex concepts, but definitely very interesting to learn from."
- "I would recommend the lessons for next year's class. Very helpful. Relates to industry which is important since we're usually stuck in academics."
- "These were great lessons. I really appreciated them."
- "I really like these practice classes. It's really useful."
- "I like these two classes. It improved my understanding of the importance of safety and know more about what real industry is like. Which can help me in my future career."
- "They were very productive and they add value to my knowledge."
- "The what-if exercises and start-up distillation was excellent. The software was harder to pick up on with the limited time we had."
- "This course is vivid with lots of videos which are quite interesting. Pro could give us more interesting things like these in the future class."
- "Excellent Work."
- "I enjoyed the what-if analysis that helped me see the consequence of a single process change to the entire unit."
- "Enjoyed seeing the practical implication of understanding these controls and how they are relevant to the industry operations. Perhaps if we could have it in a computer lab, would allow for more individuals who are able to run simulators."
- "The lessons were well organized and helped a lot."
- "Good."
- "Very informative and beneficial class. Thank you."
- "I liked the interactive nature of the lessons."

29

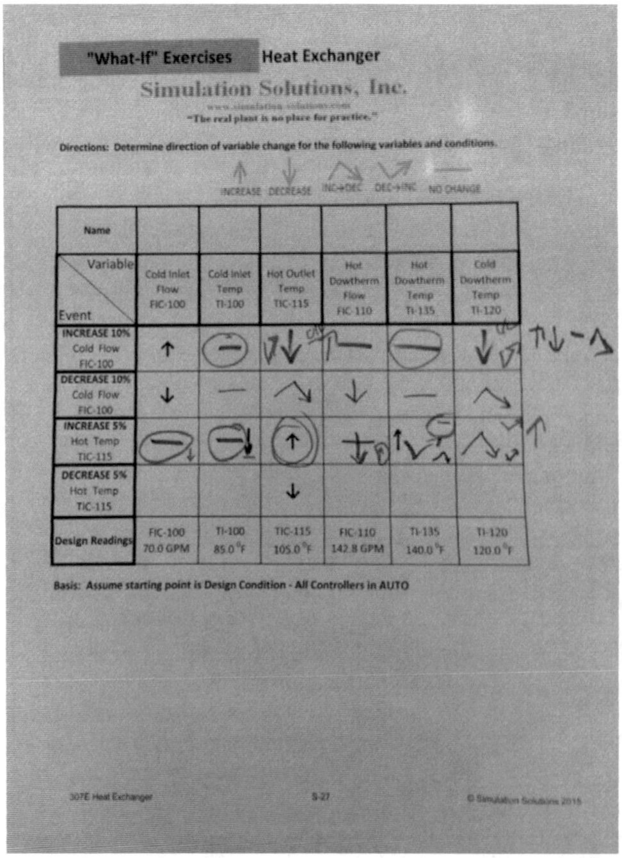

"What-If" Exercises Heat Exchanger

Simulation Solutions, Inc.

"The real plant is no place for practice."

Directions: Determine direction of variable change for the following variables and conditions.

INCREASE DECREASE INC→DEC DEC→INC NO CHANGE

Name						
Variable / Event	Cold Inlet Flow FIC-100	Cold Inlet Temp TI-100	Hot Outlet Temp TIC-115	Hot Dowtherm Flow FIC 110	Hot Dowtherm Temp TI-135	Cold Dowtherm Temp TI-120
INCREASE 10% Cold Flow FIC-100	↑	—	↓	—	—	↓
DECREASE 10% Cold Flow FIC-100	↓	—	↗↘	↓	—	↗↘
INCREASE 5% Hot Temp TIC-115	—	—	↑	↓	↑	↗↘
DECREASE 5% Hot Temp TIC-115			↓			
Design Readings	FIC-100 70.0 GPM	TI-100 85.0 °F	TIC-115 105.0 °F	FIC-110 142.8 GPM	TI-135 140.0 °F	TI-120 120.0 °F

Basis: Assume starting point is Design Condition - All Controllers in AUTO

307E Heat Exchanger S-27 © Simulation Solutions 2015

30

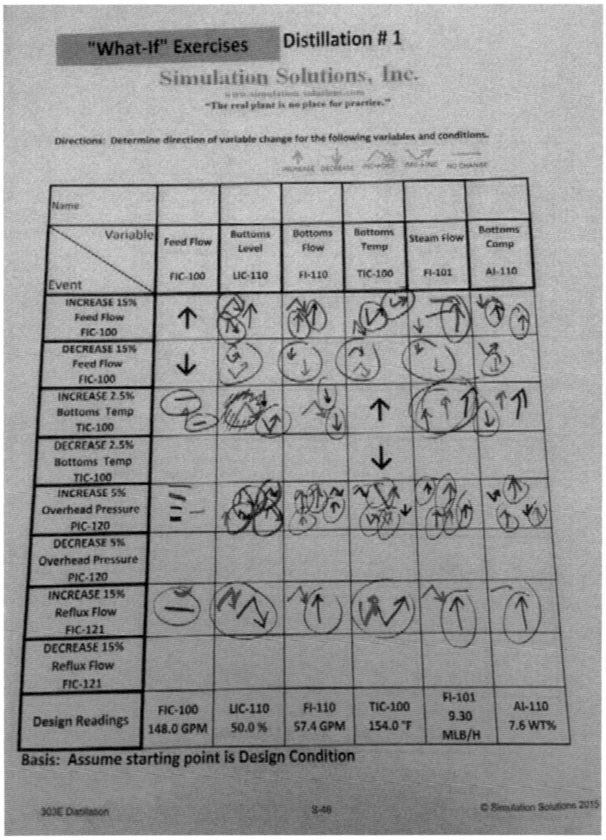

31

Heat Exchanger/Distillation “What-If” Discussion

- Students were asked to predict how Controllers and Indicators would respond to a setpoint change. (i.e. Increasing Feed to a Distillation Tower or Increasing the Temperature Controller on the Outlet of a Heat Exchanger).

- When asked the results of an increase to the Outlet Temperature Setpoint, students incorrectly predicted that both the process flow and heating material flow would affected, when only the heating material flow changes.

- Students did not always correctly predict Controller responses. For example: When more Feed is put down a Distillation Column the bottoms level will rise, but eventually come back to set point given the controller in this system. This is achieved by the Bottoms Flow increasing. Some students only predicted that this level would rise, and not return to setpoint.

32

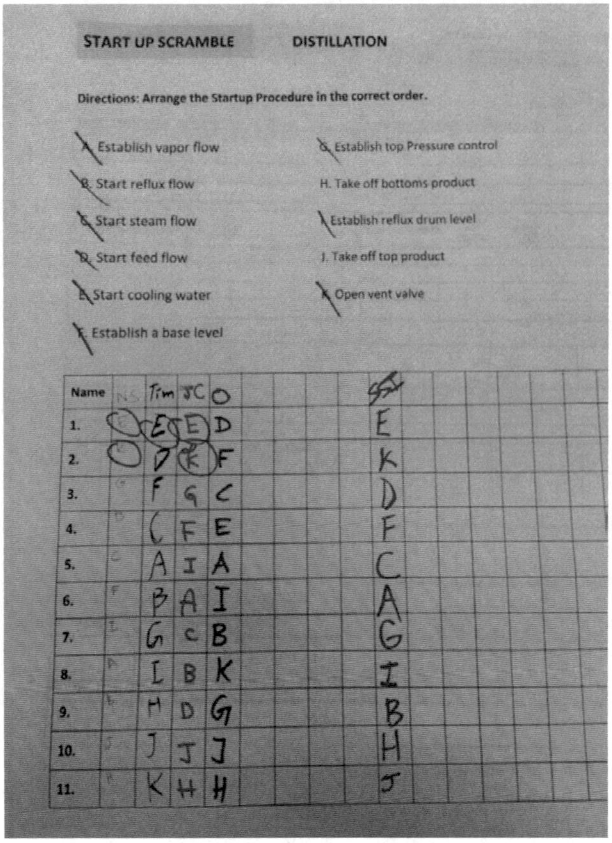
START UP SCRAMBLE DISTILLATION

Directions: Arrange the Startup Procedure in the correct order.

A. Establish vapor flow
B. Start reflux flow
C. Start steam flow
D. Start feed flow
E. Start cooling water
F. Establish a base level
G. Establish top Pressure control
H. Take off bottoms product
I. Establish reflux drum level
J. Take off top product
K. Open vent valve

Name	Tim	JC	O	[illegible]
1.	E	E	D	E
2.	D	K	F	K
3.	F	G	C	D
4.	C	F	E	F
5.	A	I	A	C
6.	B	A	I	A
7.	G	C	B	G
8.	I	B	K	I
9.	H	D	G	B
10.	J	J	J	H
11.	K	H	H	J

33

Distillation Startup Discussion

Safety Concerns

- Starting Steam Before Cooling Water (C before E)
- Starting Steam before opening Vent Valve. (C before K)
- Starting Steam Before Establishing a Base Level (C before F)

Sequence Concerns

- Establishing Overhead Vapor Flow before Starting Steam (A before C)
- Starting Reflux Flow Before Reflux Drum Level is Established (B before I)

34

Lessons Learned

<u>Lessons learned</u>

By scheduling the Project Progress Review Prior to the case study, students were lead in a direction that required them to do the case study work prior to the Progress Review. This seemed to help decrease procrastination.

Students were inspired and enjoyed opportunities to use computer software that simulated process equipment.

35

Acknowledgements

Big Thanks to Don Glaser and Matthew Garvey!

Thanks to the Columbia Students of CHEN E4501! Go Lions!

36

Conclusion and Future Work

Successful initial integration of SaChE Recommendations into curriculum.

Basic approach with Block Flow, Process Flow, and Process Control Diagrams

New approach related to operator training and education.

More process simulation in the classroom with CHEN E4500 Principles of Process Design

37

References

ASPENTECH V8.6 Software (used for design work)

Beer, David and David McMurrey, *A Guide to Writing as an Engineer 2nd Edition*. Wiley, New York, 2005

Crowl, Daniel A. and Joseph F. Louvar, *Chemical Process Safety" Fundamentals with Applications 3rd Edition*, Pearson, New York, 2011

Felder, Richard M. and Ronald W. Rousseau, *Elementary Principles of Chemical Processes 3rd Edition*, 2000, John Wiley and Sons, New York, New York. Section 4.3a Flowcharts, p90-92

Graham, Nasby, "Using Process Flowsheets as Communication Tools" *Chemical Engineering Progress*, Volume 108, No. 10.Oct 2012. p36-44, Available in Chemical Engineering Progress on line

i.ytimg.com/vi/8A1xSCUtB-M/hqdefault.jpg accessed 22 Jan 2015

Perry, Robert H. and Don W Green, *Perry's Chemical Engineering Handbook*, 7th Edition, 1997, McGraw Hill, New York, New York

Seborg, Dale E., Thomas F. Edgar, and Duncan A. Mellichamp, *Process Dynamics and Control*. Wiley Inc, New York, New York (1989) p7-9

Walker, Vornel, "Designing a Process Flowsheet" *Chemical Engineering Progress*, Volume 105, No.5. May 2009. p15-21, Available in Chemical Engineering Progress on line

www.aiche.org/academy/courses/sache/sache-certificate-program accessed 19 Jan 2015

www.aiche.org/ccps/community/students-teachers-and-communities/certificate-safety-achievement accessed 19 Jan 2015

www.bing.com/images/search?q=jacksonville%2C%20florida%20chemical%20plant%20dec%202007&qs=n&form=QBIRMH&pq=jacksonville%2C%20florida%20chemical%20plant%20dec%202007&sc=0-0&sp=-1&sk=#view=detail&id=757F70F7B398E519B65D6CACB55A59E57E50CB72&selectedIndex=5 accessed 22 Jan 2015

www.bing.com/images/search?q=pasadena%20texas%20oct%201989&qs=n&form=QBIR&pq=pasadena%20texas%20oct%201989&sc=0-19&sp=-1&sk=#view=detail&id=1A52930A92CCEC5F28D374DBCC19EA634FE9A02F&selectedIndex=1 22 Jan 2015

www.csb.gov accessed 3 Feb 2015 used for case study videos

www.csb.gov/assets/1/7/Tesoro_Anacortes_2014-May-01.pdf accessed 28 Jan 2015

www.csb.gov/assets/1/19/Final_Report_small.pdf accessed 28 Jan 2015

www.sache.org accessed 2 Jan 2015, used for SaChE

www.srai.org/wp-content/uploads/2010/06/pic2.jpg accessed 22 Jan 2015

www.toxipedia.org/download/attachments/9353/Seveso.chem.jpg?version=1&modificationDate=1238498605000&api=v2 andhttp://www.bing.com/images/search?q=Seveso+Disaster+Affected+Area&Form=IQFRDR#view=detail&id=A9CEBE175583891AADF628A0E49D369903208B4E&selectedIndex=0 accessed 22 Jan 2015

www.bing.com/images/search?q=port%20wentworth%20georgia%20feb%202008&qs=n&form=QBIR&pq=port%20wentworth%20georgia%20feb%202008&sc=0-24&sp=-1&sk=#a accessed 22 Jan 2015

www.bing.com/images/search?q=texas%20city%20mar%202005&qs=n&form=QBIR&pq=texas%20city%20mar%202005&sc=8-11&sp=-1&sk=#view=detail&id=F2A258E4AD124686AFE250ED468FBB03FA6C7066&selectedIndex=6

38

(188a) Using CO2 to Synthesize Commodity Chemicals

Stowers, K.,

Originally presented on: 11/9/2015 15:15:00 - 15:40:00

Carbon dioxide is an abundant waste product that contributes to poor atmospheric conditions. Advances have been made in using unwanted carbon dioxide for hydrogenation (C-H bond formation) and oxidation (C-O bond formation). Although the framework for carboxylation is found in carbon dioxide, C-C bond formation has been less explored. Efforts toward characterizing the principles that mediate C-C bond formation are presented. The mediation of crystallite size and metal content as well as variation in support materials have different effects on carbon dioxide activation toward C-C bond formation and toward dissociation. Synthesized catalysts are characterized for particle size, surface area, oxidation states, and reactivity using standard tools such as X-ray diffraction, physisorption, X-ray photoelectron spectroscopy, and a plug flow reactor. Ultimately, our aim is to understand how to use these principles to form carboxylic acids such as acrylic acid from carbon dioxide and ethylene.

432829 Benzomics: A High Dimensional Biology Perspective to Benzene Health Risk

Abstract:

Exposure to benzene constitutes a major environmental concern, since it is considered as an established cause of acute myeloid leukemia (AML), myelodisplastic syndromes (MDS), and probably lymphocytic leukemia and non-Hodgkin lymphoma (NHL) in humans. Even though benzene is leukemogenic at relatively low occupational levels of exposure, its association with leukemia in the general population exposed to benzene in the ppb range is somehow questionable. Identification of molecular markers of exposure and effect at real life environmental exposure levels is expected to provide additional insights on this association. It is widely accepted that the transcriptome is dynamic and that it reflects the organism's immediate and genome-wide response to environmental exposure and endogenous cues. From previous studies, it has been found that benzene exposure on the transcriptome of peripheral blood mononuclear cells (PBMCs) from a population of shoe factory workers, resulted on differentially expressed genes. A subsequent analysis of PBMCs from a population with well-characterized occupational exposure to benzene using two microarray platforms (Affimetrix and Illumina) revealed a large number of differentially expressed genes to be associated with benzene exposure. Gene pathway and ontology analysis demonstrated over-representation of genes involved in apoptosis and immune/inflammatory response. Furthermore, highly significant widespread perturbation of gene expression at all exposure levels has been observed, while, the AML pathway was among the pathways most significantly associated with benzene exposure, while a 16-gene expression signature associated with all levels of benzene exposure was identified.

The aim of the current study is to focus on the leukemogenic benzene potential for the general population, experiencing exposure to benzene at the levels of few ppb (μg/m^3) within an exposomics approach. The later relies on the evaluation of sensors technology for exposure assessment, the use of internal dosimetry and biology based dose response for risk estimation as well as the use of transcriptomcs for the assessment of exposure to effect modifiers. A major differentiation to existing studies, is that benzene is not treated alone, but co-existing to the homologous compounds toluene, ethylbenzene and xylene; this is reflecting the actual mixtures people are encountering in real environmental setttings. The variable levels of biological organization involved in this holistic view of mixture toxicology and cumulative exposure and risk assessment suggest that different technologies need to be brought to bear in order to obtain a comprehensive view of how co-exposure to multiple chemicals affects the overall phenotypic response of individuals. Technological variability introduces the need for better data integration and assimilation and for the development of novel data analysis and hypothesis generation and testing procedures in order to best elucidate the biological mechanisms underlying mixture toxicity.

This systems toxicology approach to mechanistically-based risk assessment of environmental chemical mixtures can be tackled with an integrated, multi-layer computational methodology, ideally comprising the following steps:

a) Characterization of exposure factors quantifying the parameters that affect human exposure to environmental chemicals, such as time-activity relationships, seasonal and climatic variation, and consumer choice. These exposure factors can be used to derive aggregate and cumulative exposure models, leading in probabilistic exposure assessments. Aggregation can be done across exposure pathways and routes and even across different exposure scenarios, if the relevant exposure metric or the imputable biological or physiological effect can be related to these scenarios. For instance, exposure to volatile organic compounds (VOCs) such as benzene or toluene and mixtures thereof may occur both from environmental media and in specific occupational settings. A cumulative exposure scenario for these substances would have to take stock of the actual variability of exposure across these different settings throughout typical days for the same period in an individual's lifespan.

b) Current toxicological state of the art combines estimations of biologically effective dose with early biological events to derive dose-effect models, which can be used in combination with the probabilistic exposure estimates to derive biomarkers of exposure and/or effect. Combined use of epidemiological, clinical and genetic analysis data may shed light on the effect of risk modifying factors such as lifestyle choices and DNA polymorphisms. Observation of real clinical data and /or results of biomonitoring, if coupled with the exposure/effect biomarker discovery systems, can produce biomarkers of individual susceptibility and thus allow estimations of individual response to toxic insults. Transcriptomics, are key technology to this kind of analytical and data interpretation process.

c) The analysis of the biomarker data (including results on biomarkers of exposure, effects and individual susceptibility) results in the integrated assessment of risk factors. Use of information on risk factors with molecular dosimetry data (i.e. estimation of the actual internal and biologically effective dose of xenobiotic substance found in the target organ and, indeed, perturbing cellular response) enables population risk studies to be done, by converting generic exposure profiles into population risk metrics having taken into account inter-individual variability of response and exposure uncertainty.

Environmental and personal exposure to benzene were retrieved from a series of studies carried out by the authors, including several environmental settings and population groups, occupational or not. Internal exposure to benzene and its toxic metabolites was assessed through a Physiology Based BioKinetic (PBBK) model, for a quaternary mixture of BTEX. To translate the estimated internal dose (concentration of toxic metabolites in the bone marrow) into actual risk estimates, a biology-based dose- response model (BBDR) that links external exposure to a specific end-point, which, in the case of benzene, is primarily the risk of leukemia was developed. In this work a method was applied, based on the decomposition of the dose-response relationship into a set of causal micro-relations each one describing a biologic process separately. Instead of evaluating the relationship between administered dose and cancer risk "directly" through an empirical-statistical model, here this relationship is decomposed into two different parts: the first one links the administered dose to the total amount of metabolites produced (internal dose), while the second one connects the internal dose to the probability of cancer. The statistical relation between internal dose and cancer probability was calculated parameterizing function as given by Crump and Allen. In particular the administered dose was calculated assuming an average person of 70 kg (adult) who inhales 10 $m^3$of air in 8 hours for an occupation period of 40 years over a life of 70 years.

Transcriptomics analysis showed that among the most important molecular pathways differentially expressed were the ones regarding inflammation mediated by chemokine and cytokine signaling, apoptosis as well as oxidative stress. The extent of modulation was proportional to the time of exposure, and to the increase of the benzene:toluene ratio; At short exposure regimes, low concentrations seem to produce hormesis. At higher exposure regimes, the activation of homeostasis results seem to be counterbalanced, thus responses are dose-dependent It has been suggested that chemically-induced chronic inflammation, apoptosis and oxidative stress may lead to carcinogenesis; the latter is one of the most important adverse health effects of benzene. Pathway analysis of the results mentioned above indicated that the apoptosis signaling pathway is activated under the co-exposure to the BTEX chemical mixture. Using quantitative PCR analysis the extent of modulation in the expression of cytochrome P450 (CyP450) as a function of exposure to BTEX mixtures with variable toluene:benzene ratio was quantitatively investigated; CyP450 is generally known to regulate benzene metabolism. They are time-dependent (i.e., several differences can be observed in gene expression modulation after 4 and 24 hours). In general, the toluene-richer mixture induces suppression of the difference in gene expression compared to the controls for practically all the CyP450 gene family. The phenotypic outcome of the CyP450 suppression would be reduction of its catalytic activity in VOC metabolism. This would support the metabolic inhibition hypothesis for the interaction between the constituents of the quaternary BTEX mixture. Overall, the toxicogenomics results would suggest that co-exposure to BTEX reduces benzene metabolism.

Based on the above exposure measurements and translating them into internal exposure, it was estimated that general population of the city runs an average lifetime leukemia risk (due to benzene exposure) equal to $4.1 \cdot 10^{-5}$ ($2.1 \cdot 10^{-5}$ to $8.2 \cdot 10^{-5}$), while for the occupational groups the estimated risk is as high as $11 \cdot 10^{-5}$. The leukemia risk distributions derived herein, represent the variation in activity patterns that result in different levels of exposure and the variability of the population physiological response to environmental stressors. Benzene toxicity is intimately tied to its complex metabolism and distribution. Key enzymes involved in the metabolism of benzene, including CYP2E1, the quinone reductase NQO1, and myeloperoxidase are polymorphic. For example, a 7.6-fold difference in benzene-induced hematotoxicity in workers was observed among gene variants of CYP2E1 and NQO1. Thus the expression of genetic polymorphisms may modulate the susceptibility of an individual or ethnic group to the effects of benzene exposure.

(188c) Capturing Carbon Dioxide with a Green Focus

Duarte, P. D.,
Gottlund, S. S.,

Originally presented on: 11/9/2015 16:05:00 - 16:30:00

A unique collaboration between the Air Products and Chemicals Inc. (APCI) and the US Department of Energy (DOE) has resulted in the design and implementation of an innovative process to capture CO2 at the APCI Port Arthur Facility. The CO2 is captured, via a Vacuum Swing Adsorption process, from a Syngas stream generated in conjunction with the production of hydrogen by means of steam methane reforming. The captured CO2 is purified, dried, compressed, and transported via pipeline to Denbury Resources where it is utilized in the process of Enhanced Oil Recovery (EOR). As of April 2014 over 1.35MM tonnes of CO2 have been delivered to the customer from the APCI Port Arthur facility.

437680 Optimizing Health, Safety, and Environmental Sustainability." Mayan Pigments: Organic/Inorganic Hybrid Materials - Art, Science, Cancer

Abstract:

The Ancient Maya developed specialized pigments and produced a large quantity of magnificent murals and installations, during their classic period (AD 250 – 900). The most well known Mayan pigment is Azul Maya, (Maya Blue); which was used as a sacrificial paint on victims. It is very stable and virtually indestructible and has been termed, "one of the great technological and artistic achievements of Mesoamerica".1 We have developed a complete understanding of Mayan pigments and have a company that produces Mayan Pigments for many applications.2 Mayan Pigments Inc. produces the materials (hybrid organic/inorganic compounds) consisting of an indigo derivative and Palygorskite, a common clay called Sacalum (Sacred Earth in Mayan) by the ancient Maya. The science is a detailed study of the manufacture and structure of Azul Maya, which has led to a series of new materials with an "Infinite Color Space".3 Finally, we have discovered that the Mayan Pigments are very effective against various cancers in cell-based assays.4 This discovery is now under evaluation for human use. This discovery was fostered by a study of the ancient medicinal uses of indigo. A recent British article describes the use of indigo based Celtic war paint as possibly a cure for cancer.5 1. Field Museum, Centuries-old Maya Blue Mystery Finally Solved, Science Daily, February 28, 2008. 2. Polette, L.A., Ugarte, N., Yacaman, M.J., and Chianelli, R.R., Maya Blue, Discovering Archeology, 46-53, August 2000. 3. Polette-Niewold, L.A., Manciu, F.S., Torres, B., Alvarado Jr., M., and Chianelli, R.R., Organic/Inorganic Complex Pigments: Ancient Colors Maya Blue, J. Inorganic Biochemistry, 101, 1958-1973(2007). 4. Das, S. and Chianelli, R.R., Treatment of Cancer with Complex OrganicInorganic Pigment Compositions, 8,563,595 October 22, 2013. 5. Hall, C., Celts' warpaint may be weapon to beat cancer, The Telegraph, August 14, 2014, http://www.telegraph.co.uk/health/expathealth/4201066/Celtswarpaint-may-be-weapon-to-beat-cancer.html, August 15

408516 Flash Point of Petroleum Fuels: Evaluation and Predictive Methods

Abstract:

Flash Point of Petroleum Fuels: Evaluation and Predictive Methods

Sara S. AlQaheem and M. R. Riazi*

Department of Chemical Engineering, Kuwait University, Kuwait

* Speaker and corresponding author. Professor and Head / AIChE Fellow (www.RiaziM.com)

Abstract:

Flash point is an important safety data for petroleum fuels and derived products during processing, transportation and storage of these materials especially in a high temperature environment [1, 2].

In the first phase of this work, data on flash point of 65 pure hydrocarbons and 75 petroleum mixtures and coal liquid samples were collected and used to evaluate ten most widely used predictive methods available in the literature. Errors varied in the range of 1.5 to 5% for pure hydrocarbons and 4 to 11% for the petroleum mixtures.

In the second phase of this work, based on the method suggested by Butler and Cooke [3] that the product of molecular weight and vapor pressure at the flash point is a constant for any hydrocarbon, a very simple relation was derived for the flash point of hydrocarbon systems. Relations between vapor pressure and molecular weight of hydrocarbon systems with their boiling point were used to show that the flash temperature occurs at 70% of boiling point in absolute degrees. Evaluations for both pure compounds and mixtures showed that the simple derived relation is better or at least equivalent to the existing methods while boiling point is used as the only available data. This simplified method predicts the flash point with an error of 1.7% for pure and 2.8% for hydrocarbon mixtures.

References:

[1] Riazi, M. R. (2005). *Characterization and Properties of Petroleum Fractions*. West Conshohocken, PA: ASTM International.

[2] Riazi, M. R., Eser, S., Agrawal, S., Pena Diez, J.L. (2013). *Petroleum Refining and Natural Gas Processing*. West Conshohocken, PA: ASTM International.

[3] Butler, R. M., and Cooke, G. M. (1956). Prediction of Flash Points of Middle Distillates. *Industrial and Engineering Chemistry*, Vol. 48, No. 4, 808-812.

15AIChE
Annual Meeting, Salt Lake City, UT

CFD Modeling of Soot Emission and Flare Efficiencies for Controlled Flare Tests

Vijaya Damodara[1],
Daniel H. Chen[1], Helen H. Lou[1], Hashim Almrayani[1], Ajit Patki[2], Arokiaraj Alphones[1], Xianchang Li[2], Christopher B. Martin[3] and Matthew R. Johnson[4]

[1]Dan F. Smith Department of Chemical Engineering, Lamar University, Beaumont, TX, USA
[2]Mechanical Engineering, Lamar University, Beaumont, TX, USA
[3]Chemistry and Biochemistry, Lamar University, Beaumont, TX, USA
[4]Energy & Emissions Research Lab, Carleton University, Ottawa, ON, Canada

15AIChE
Annual Meeting, Salt Lake City, UT

Outline

- Introduction
- Reduced combustion mechanism LU3.0.1
- Results and validation
- Modeling of flares using ANSYS Fluent®
- Results and discussion
- Conclusion

Introduction

- Flare emissions account for 60% of the Highly Reactive Volatile Organic Compounds (HRVOCs) based on the 2007 HRVOC special emissions inventory
- Over 95 percent of the waste gases flared are from oil/natural gas production and olefin/polymer plants.
- Several intermediate products are formed, some of which plus unburned hydrocarbons escape as emissions
- Recent studies/gas imaging IR camera suggest that flares may not be as efficient as claimed
- Many combustion mechanisms are available readily but are quite large involving 1000's of reactions
- Smaller versions of these mechanisms were developed to be applicable in CFD simulations

Combustion products and intermediate species in flares

- Emissions from flaring constitute unburned hydrocarbons, CO, soot, and other partially burned and altered hydrocarbons along with CO2 and water
- Rigorous combustion mechanisms suggest numerous incomplete combustion species are produced in combustion reactions
- These include radicals (HO, & HO_2), radical producing species (CH_2O, C_2H_4O, HONO), NO_x, CO, and soot
- In recent past, Lamar research team has simulated several laboratory flames (e.g., Sandia flame and McKenna flat flame) with good agreements
- Soot or visible smoke is of particular concern for flare operators / regulatory agencies

Full/Reduced Combustion Mechanisms

15AIChE Annual Meeting, Salt Lake City, UT

- GRI-Mech 3.0 - (53 species and 325 reactions)
- USC-II - 111 species and 784 reactions. Based on GRI -3.0
- Lamar research group developed 50-species reduced mechanisms for the combustion of C_1-C_3 light hydrocarbons (LU1.0, 1.1, and 2.0 based on GRI-3.0 and USC-I mechanisms)
- Successfully validated against experimental data in CHEMKIN-PRO
- Tested in simulations for its further validation
- LU 3.0.1 - Capable of predicting soot in TCEQ air cases in CFD simulations

Species Selection for Developing a Reduced Mechanism

15AIChE Annual Meeting, Salt Lake City, UT

- Mechanisms used: LU1.1 and LU2.0* mechanisms & USC II
- Common VOC species unchanged in new mechanism
- New reduced combustion mechanism with 50 species and 310 reactions was developed from the USC-II mechanism
- Contains important soot precursor species and can handle combustion of C_1-C_4 hydrocarbons (n-butane, 1-butene and 1,3-butadiene included)
- Predicts both volatile organic compounds (VOCs) and soot emissions
- New mechanism has been compared to the full USC II mechanism and successfully validated against experimental data in CHEMKIN-PRO

Reaction Path Diagram

CH_4-CO_2 with 15 species

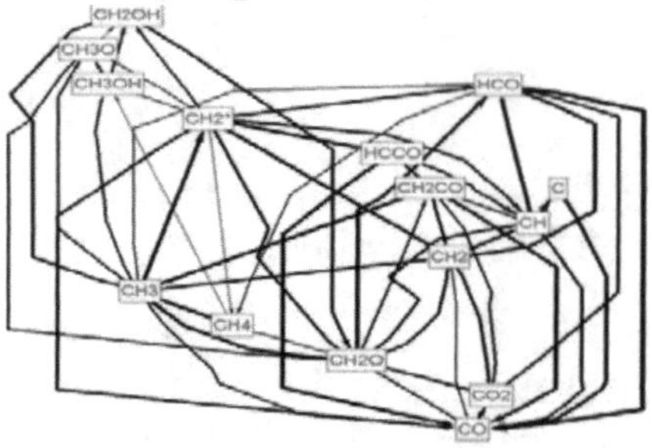

C_3H_6-CO_2 with 15 species

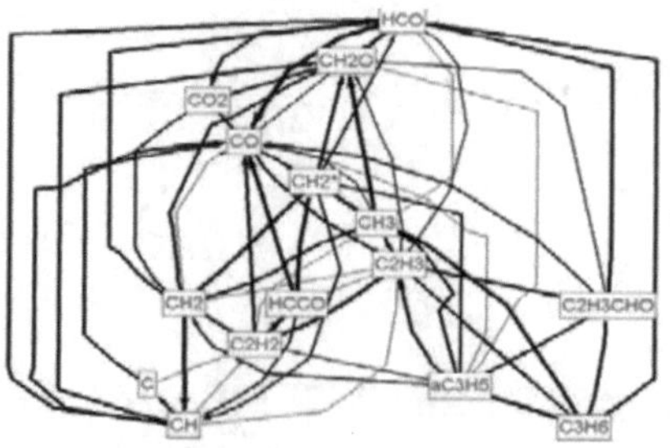

C_4H_{10}-CO_2 with 15 species

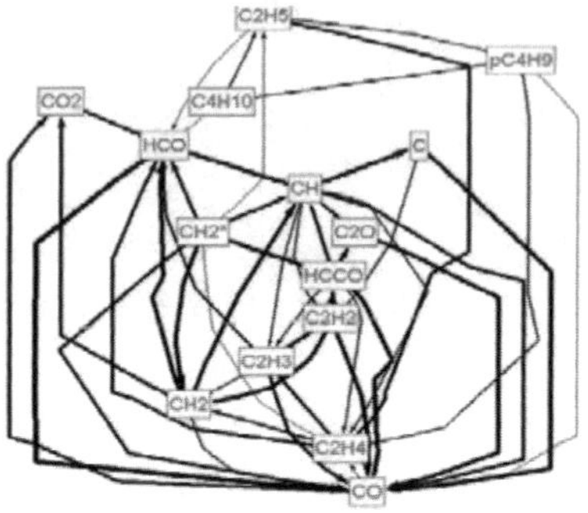

C_2H_6-CO_2 with 15 species

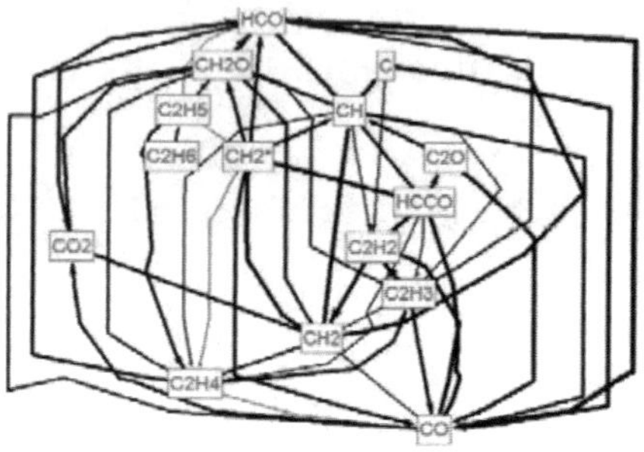

Validation:
Key Performance Indicators of a Flame

- **Laminar flame speed**
  - Speed at which a laminar flame propagates through a pre-mixture of fuel and air
- **The ignition delay time**
  - Period between the creation of a combustible mixture when the fuel is injected in an oxidizing environment, and the sustained, onset of the rapid reaction phase leads to the rise of temperature and pressure
- **Adiabatic flame temperature**
  - Measure of the maximum temperature that could be reached by combusting a particular gas mixture under a specific set of conditions

15AIChE Annual Meeting, Salt Lake City, UT

Validation with Experimental Data

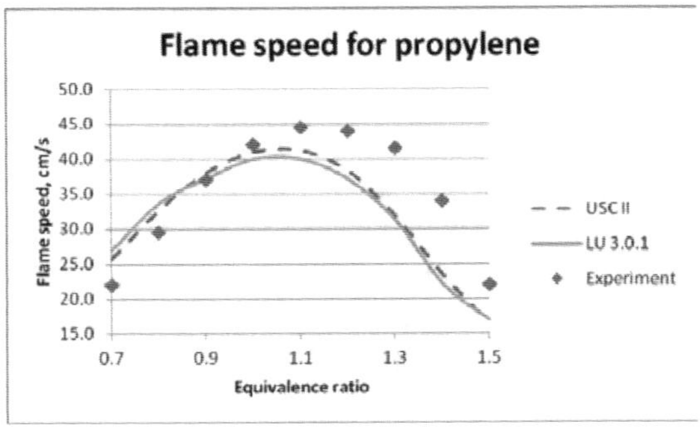

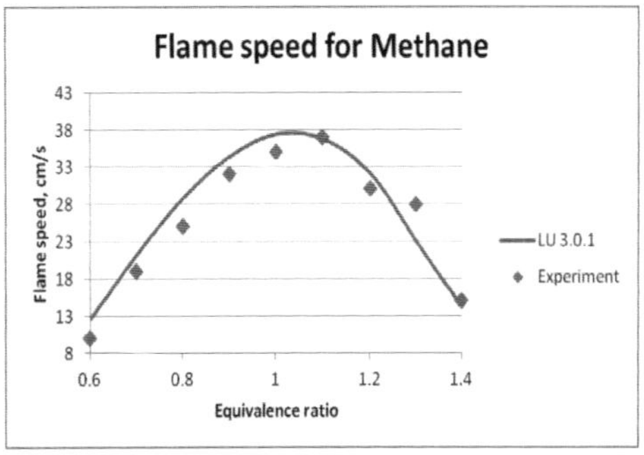

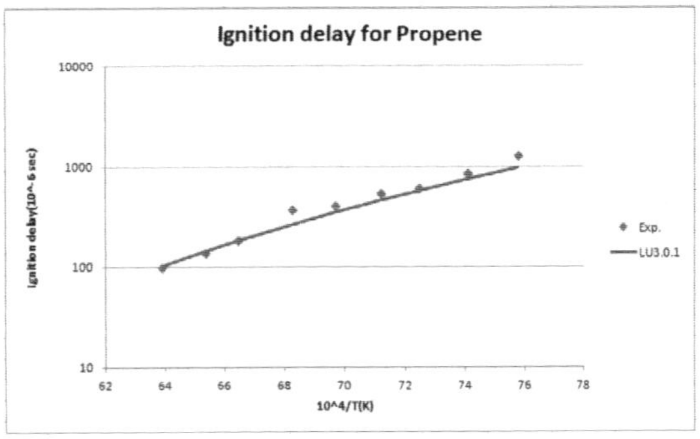

15AIChE Annual Meeting, Salt Lake City, UT

CFD modeling

Prediction of fluid flows by means of

Mathematical modeling (partial differential equations)

⇩

Numerical methods (discretization and solution techniques)

⇩

Software tools (solvers, pre- and postprocessing utilities)

Compared to experimental study Cheap(er) Fast(er) Parallel Multiple-purpose

Process flow diagram of CFD modeling

Problem Statement ⟶ Mathematical Model ⟶ Geometry and Mesh Generation

↓

Postprocessing ⟵ Run the Simulation ⟵ Solver Settings

↘ Data Analysis

15AIChE
Annual Meeting, Salt Lake City, UT

Air-Assisted Flare: Sep. 2010 TCEQ/John Zink Flare Campaign, Tulsa, OK

Elevated Position

2010 TCEQ Flare Study Final Report, UT/CEER, TCEQ PGA No. 582-8-86245-FY09-04 and Task Order No. UTA10-000924-LOAT-RP9, Aug. 1, 2011.

15AIChE
Annual Meeting, Salt Lake City, UT

TCEQ Air assisted flare stack and flare tip

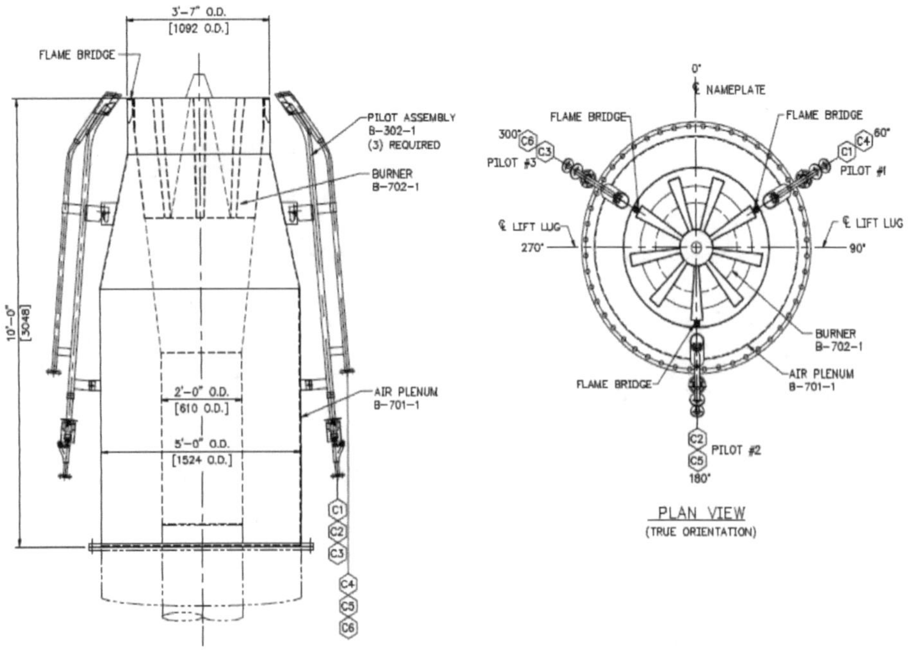

15AIChE
Annual Meeting, Salt Lake City, U

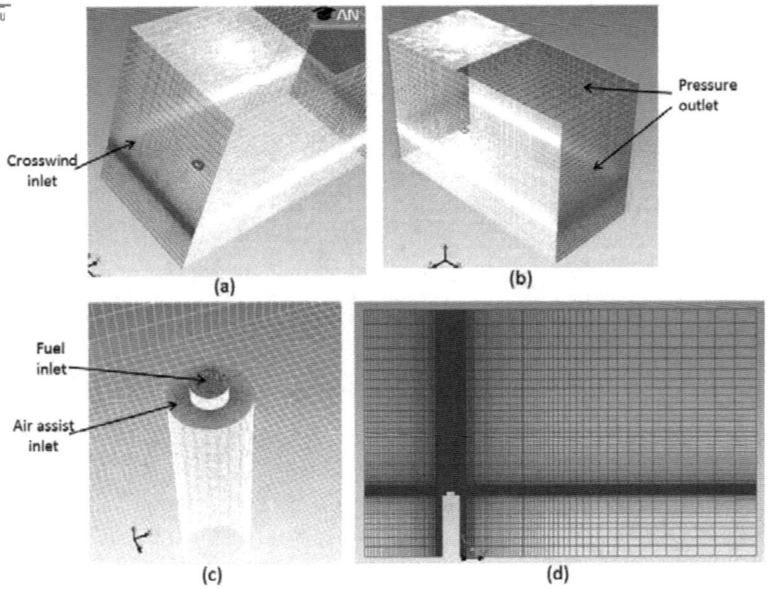

Orthogonal quality of grid =0.76
grid size = 0.8 million

15AIChE
Annual Meeting, Salt Lake City, UT

Comparison: EDC and PDF Turbulence-Chemistry Models

EDC (Eddy Dissipation Concept)	PDF (Probability Density Function)
Reactions taking place in the flame are governed by the Arrhenius rates	Reactions are governed by a conserved scalar quantity known as mixture fraction
Incorporates detailed chemical mechanisms	Not so rigorous
Molar concentrations are derived from Reaction rates, which are calculated using ISAT algorithm	Molar concentrations are derived from the predicted mixture fraction fields
Any number of inlet streams can be defined	Only two inlet streams are allowed i.e. Fuel and Oxidizer
Computationally very expensive; requires 5-6 days for convergence	Requires less time for convergence; only 2-3 days

15AIChE
Annual Meeting, Salt Lake City, UT

Input Conditions for Simulated Air-Assist Cases

Case no.	Total (kg/s)	Temp (°F)	Actual Vent gas LHV (Btu/scf)	Exit velocity (m/s)	Air flow rate (kg/s)	Crosswind (m/s)
A1.1	0.1158	87	2108	0.43	18.8	5.68
A2.1	0.0446	94	2120	0.15	10.54	5.81
A3.1	0.0883	89.3	339	0.58	2.3	5.99
A4.1	0.1181	68.6	584	0.58	6.52	4.06
A5.1	0.0448	56.4	344	0.24	1.1	1.2
A6.1	0.0442	85.9	584	0.21	1.43	7.1

15AIChE
Annual Meeting, Salt Lake City, UT

CFD Modeling: PDF Results with LU 3.0.1 (2010 Tulsa Tests)

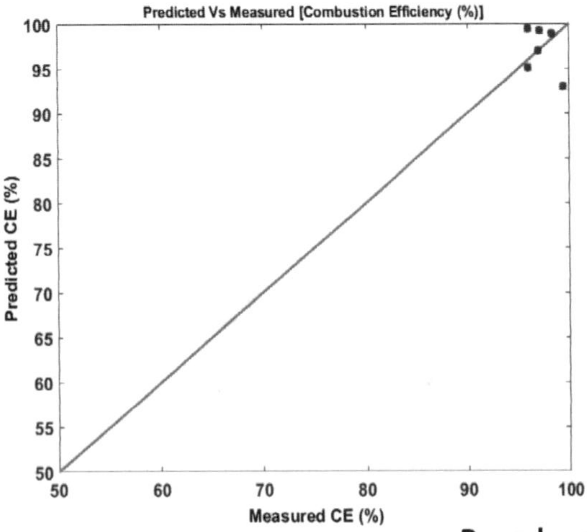

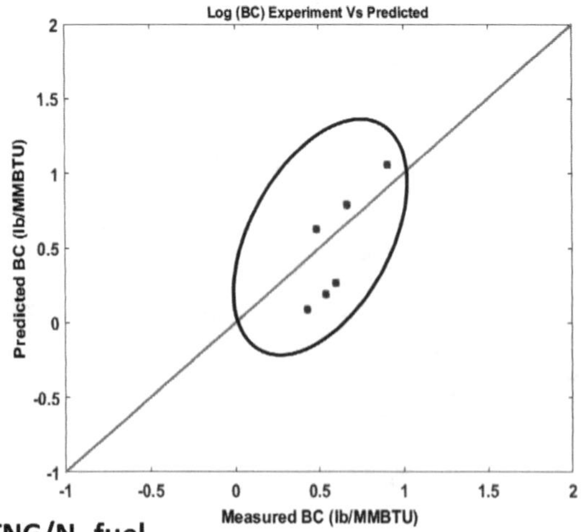

Propylene/TNG/N_2 fuel

- CE predictions: max. 6%, with an avg. of 2%
- Soot yield predictions: in the right order of magnitude, with in a factor of 2
- But PDF predicts virtually no VOCs (emissions are C and CO)

15AIChE
Annual Meeting, Salt Lake City, UT

CFD Modeling: EDC Results with LU 3.0.1 (2010 air assist Tulsa Tests)

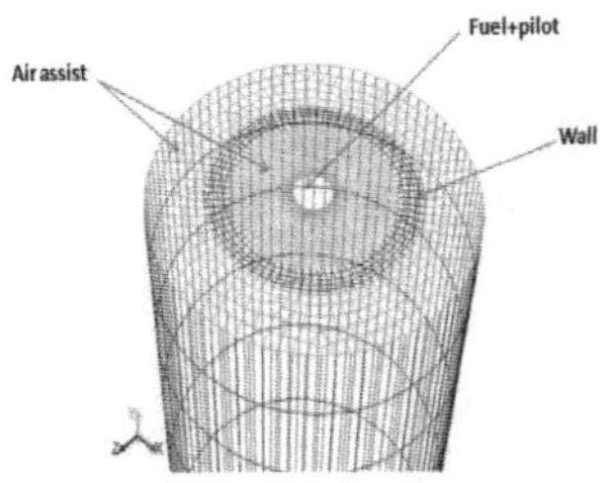

Side view of air-assisted stack showing the inlet surfaces.

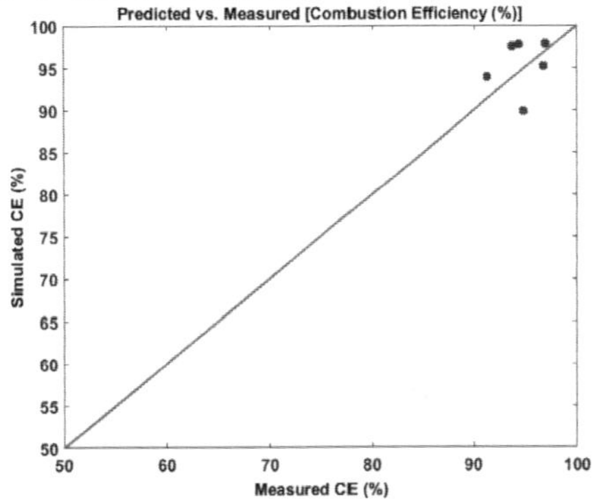

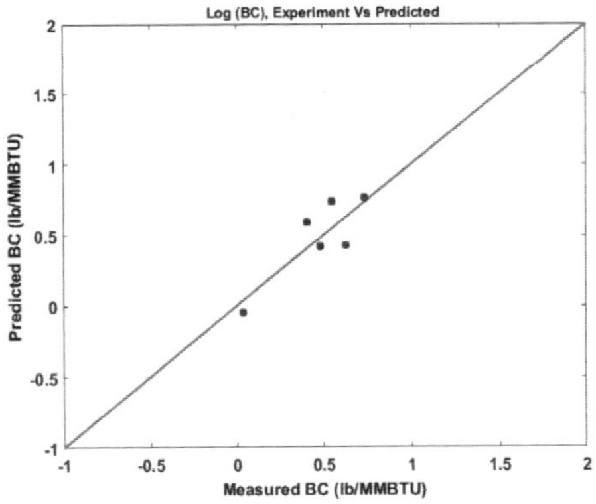

- CE predictions: Comparable to PDF, but may lead to convergence difficulties at times
- Soot yield predictions: In the right order of magnitude, also within a factor of 2

15AIChE
Annual Meeting, Salt Lake City, UT

CFD Modeling: PDF Results with LU 3.0.1 (2010 Tulsa Steam Tests)

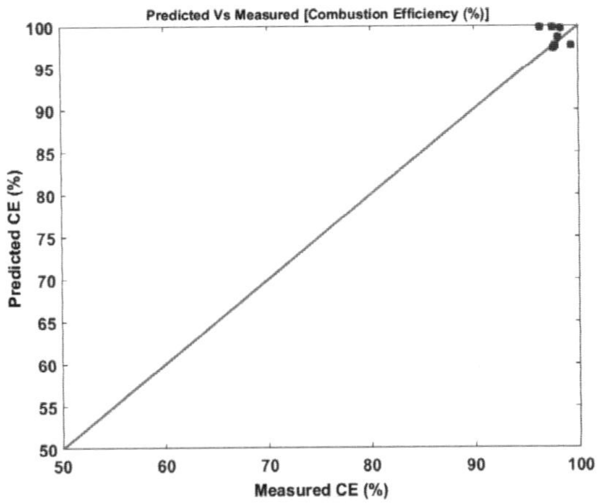

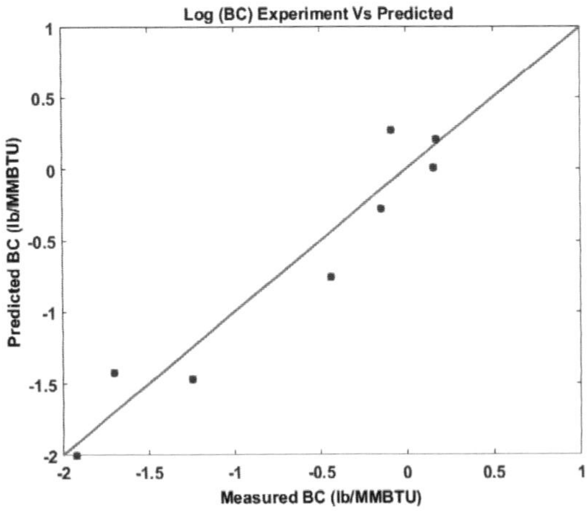

Non-Assisted Lab-Scale Flare Experiments at Energy & Emissions Research Lab, Carleton University, Ottawa, Canada

15AIChE Annual Meeting, Salt Lake City, UT

Photo showing sampling hood and flare at the Energy & Emissions Research Lab. Flare Facility at Carleton University

Species	Purity (%)	Average 6-Mix (%)	Heavy 4-Mix (%)	Ethylene (%)
Methane	99.0	85.24	74.54	-
Ethane	99.0	7.06	15.47	-
Propane	99.0	3.11	6.83	-
Butane	99.0	1.44	3.16	-
Carbon Dioxide	99.0	1.91	-	-
Nitrogen	99.999	1.24	-	-
Ethylene	99.5	-	-	100

A turbulent diffusion non-premixed flame burning hydrocarbon mixtures relevant to low-momentum flares ubiquitous in the oil and gas industry

Soot Results: PDF
Effect of C_oxid correction

15AIChE Annual Meeting, Salt Lake City, UT

6-mix (50.8 mm) Test no	Jet Velocity m/s	Soot yield simulation kg/kg	Soot yield experimental kg/kg	Experimental uncertainty (+/-)	
5	0.28	2.39E-04	5.31E-04	1.40E-04	1.40E-04
6	0.56	3.65E-04	5.32E-04	1.12E-04	1.12E-04
7	1	6.08E-04	5.96E-04	1.25E-04	1.25E-04
8	1.67	9.47E-04	3.92E-04	8.25E-05	8.25E-05
9	2.42	1.08E-03	3.56E-04	7.48E-05	7.48E-05

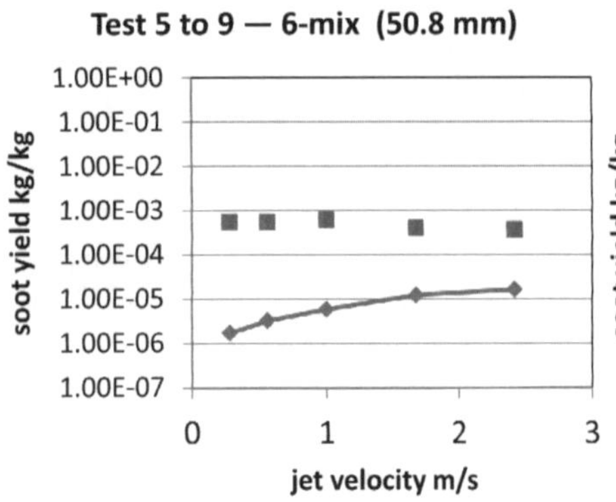

With C_oxid = 1 (default)

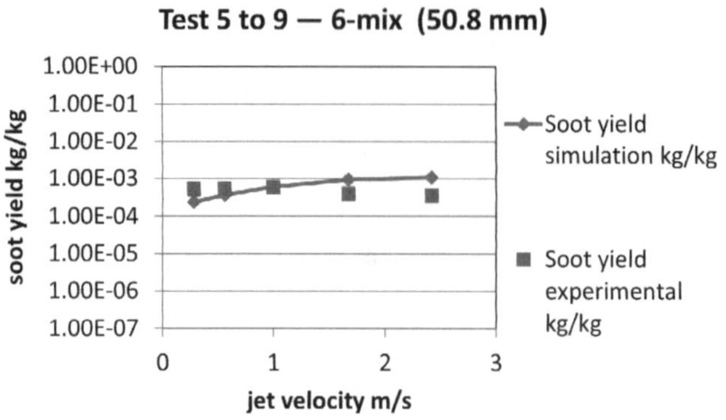

With C_oxid = 0.6

15AIChE
Annual Meeting, Salt Lake City, UT

Carleton university flare test cases-Results-NO_x and soot-PDF results

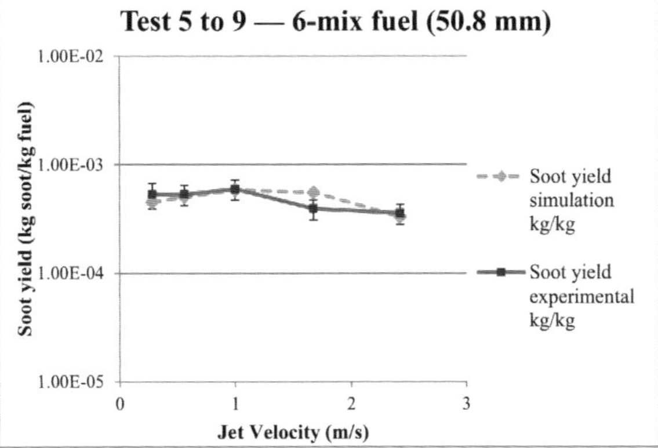

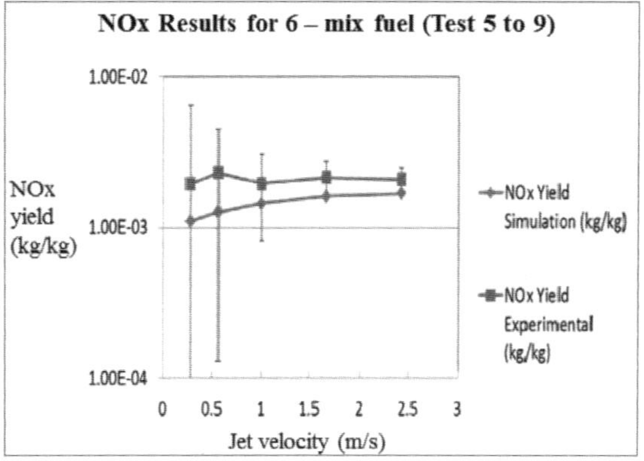

Soot /NOx Yield prediction for Carleton University 2014 Experimental Data with PDF and LU3.0.1

- For high LHV/high flow flares, simple PDF model can predict both CE/soot/NOx well .
- In the case of Carleton University's lab-scale flares, both CE and soot yield can be predicted within experimental uncertainty with LU3.0.1 and PDF approach.
- Soot yield predictions in the right order of magnitude, within a factor of 2, but virtually no VOCs emissions, only unburned C and CO emitted.

15AIChE
Annual Meeting, Salt Lake City, UT

Soot Yield Analysis-EDC

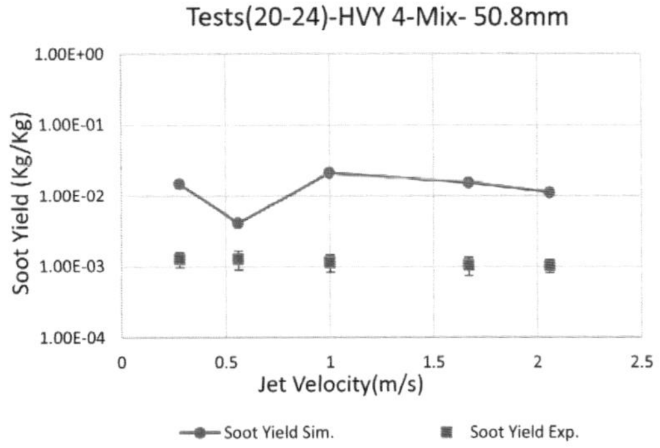

Soot Yield for Heavy 4-Mixture Fuel Test Cases (50.8 mm Jet Diameter)

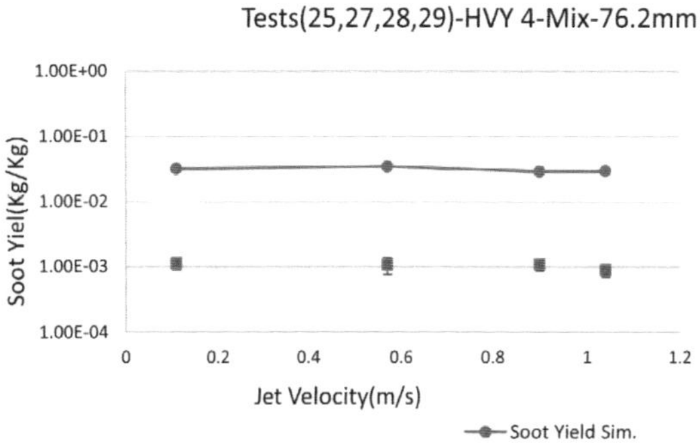

Soot Yield for Heavy 4-Mixture Fuel Test Cases (76.2 mm Jet Diameter)

Soot Yield based on the EDC Turbulence Chemistry model & Fennimore-Jones Soot Oxidation Model :

❖over-predicted for methane-based flares

Soot Yield Results for different Fuels-EDC (50.8 mm Jet-Diameter)

15AIChE
Annual Meeting, Salt Lake City, UT

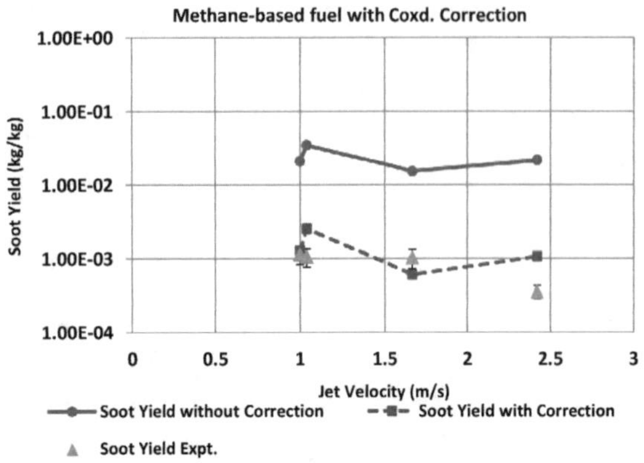

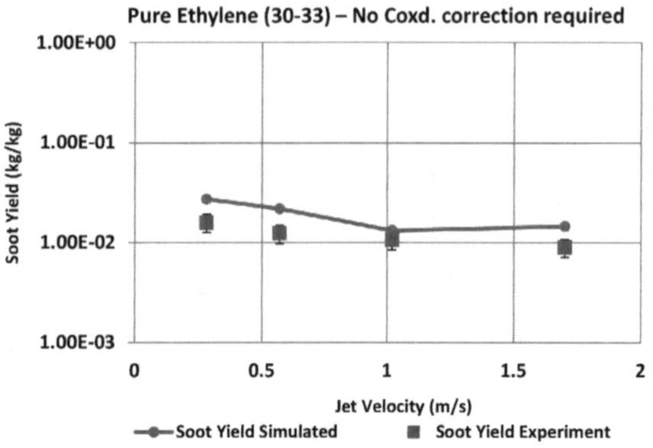

Soot Yield based on the EDC Turbulence Chemistry model & Fennimore-Jones Soot Oxidation Model :

❖ in agreement with experiments for ethylene flares

15AIChE
Annual Meeting, Salt Lake City, UT

EDC Results- Carleton University 2014 Tests

CO_2 Yield Results for 3 Different Fuel (50.8 mm Jet-Diameter)

AVG 6-Mix. Fuel (Tests6-9) HVY 4-Mix. Fuel(20-24) Pure Ethylene Fuel(30-33)

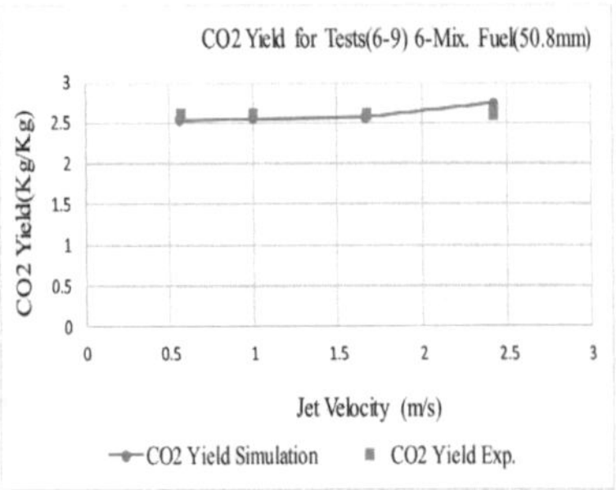

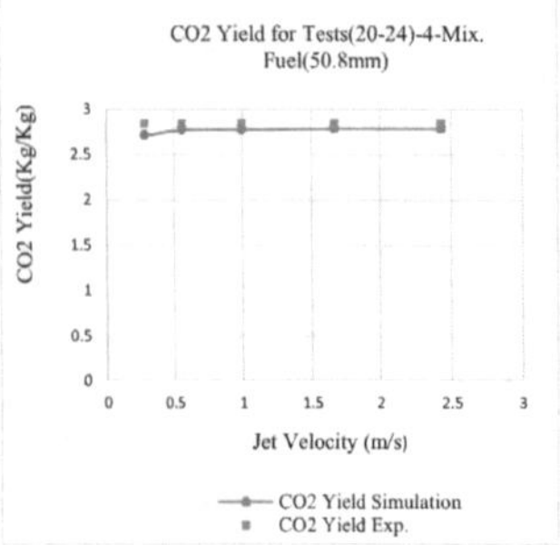

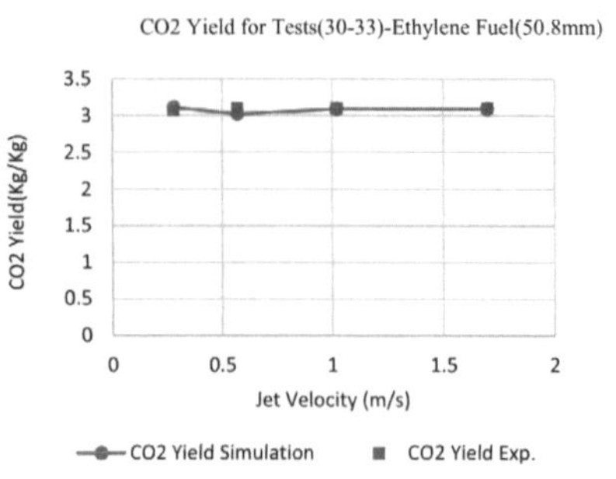

CO_2 Yield increases with jet velocity

15AIChE Annual Meeting, Salt Lake City, UT

Conclusion

PDF Model	EDC Model
CE predictions: max. 6%, with an avg. of 2% DRE is always over-predicted	Under-predicts both DRE & CE but in good agreement
Easy to converge	Convergence difficulties at times
Virtually predicts no VOCs	Reasonably predicts VOCs depending on the mechanism used
Soot Yield: ❖ Predicted within uncertainty limits after correction equation for CU experimental data ❖ Predicted in the right order of magnitude, with in a factor of 2 for TCEQ test cases	Soot yield: ❖ For CU data, o Over-predicted for methane-based flares o In agreement with experiments for ethylene flares ❖ Predicted in the right order of magnitude, within a factor of 2 for TCEQ test cases
Temperature ranges from [1760-1820]K methane-based flares and [1935-1990] K for ethylene flares	Temperature ranges from [2181-2226] K for methane-based flares range and [2378-2395]K for ethylene flares.
Lee Soot Oxidation Model works better.	Fennimore-Jones Soot Oxidation Model works better.

LU3.0.1 Reduced mechanism

- Predicts VOCs and soot
- Handles C1-C4 Species
- EDC model tends to over-predict VOCs
- Convergence difficulty in case of lower LHV flares
- Based on USC II mechanism

15AIChE Annual Meeting, Salt Lake City, UT

CFD lab at Lamar University

- **Cutting Edge High Performance Computing (HPC) Cluster**
  - 3 Thinkmate HPX XS8-2460 servers, Intel C602 Chipset - Dual 1-Gigabit Ethernet - 4U/Tower GPU, 2 x Six-Core Intel Xeon Processor E5-2630 v2 2.60GHz 15MB Cache (80W), 8 x 8GB PC3-12800 1600Mhz DDR3 ECC Registered DIMM
  - 3 X 12 core Dell servers; Intel Xeon X5670 @2.93GHz, More than 50 high speed processors, Up to 10GBs/second of data transfer speed for faster parallel computing
  - 3x24 Core Dell servers; Intel Xeon E5-2630V2 @2.6 GHz

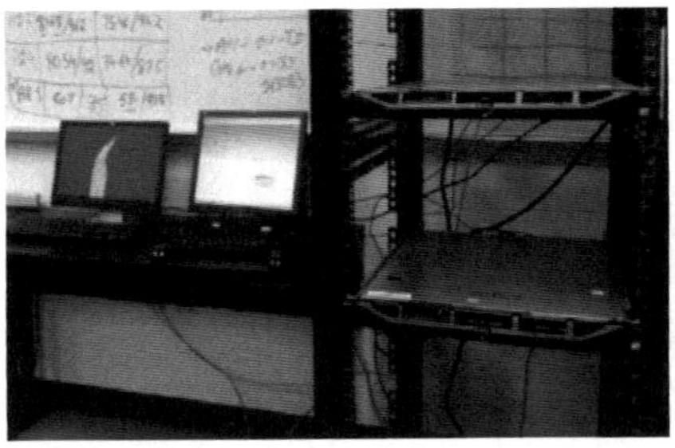

15AIChE
Annual Meeting, Salt Lake City, UT

References

- McDaniel, M., Flare Efficiency study, EPA-600/2-83-052, July 1983.
- Pohl, J., R. Payne, and J. Lee. 1984. Evaluation of the efficiency of industrial flares: Test Results. EPA-600/2-84-095. Prepared for U.S. EPA Office of Research and Development by Energy and Environmental Research Corporation (May).
- EPA. (2014). Method 9-Visual Determination of the Opacity Of Emissions From Stationary Sources, "Environmental Protection Agency, 40 CFR Ch. I (7–1–09 Edition), Pt. 60, App. A–4, Meth.9. http://www.epa.gov/ttn/emc/methods/method9.html.
- EPA. (1993).Visible Emissions Field Manual, EPA Methods 9 and 22, US. ENVIRONMENTAL PROTECTION AGENCY, Stationary Source Compliance Division, Office of Air Quality Planning and Standards, Washington, DC 20460.
- D. Allen and V. Torres, 2010 TCEQ Flare Study Final Report & Appendices," The University of Texas at Austin, TCEQ PGA No. 582-8-86245-FY09-04 and Task Order No. UTA10-000924-LOAT-RP9, Aug. 1, 2011.
- D. Corbin & M. Johnson. "Summary of Flare Efficiency and Soot Emission Rate Measurement Results at Carleton University Energy & Emissions Research Lab", TCEQ GAD Project number 582-10-94307-FY14-06, PI, D. Chen, Aug. 2014.
- K. Singh, P. Gangadharan, D. Chen, H. Lou, X. Li & P. Richmond (2014), "Computational fluid dynamics modeling of laboratory flames and an industrial flare", Journal of the Air & Waste Management Association, 64:11, 1328-1340.
- K. Singh, T. Dabade, H. Vaid, P. Gangadharan, D. Chen, H. Lou, X. Li, K. Li, C. Martin, "Computational Fluid Dynamics Modeling of Industrial Flares Operated in Stand-By Mode," Industrial Flares special issue, Ind. & Eng. Chem. Research, 51 (39), 12611-12620, , October, 2012.

15AIChE
Annual Meeting, Salt Lake City, UT

Acknowledgements

Financial support from TCEQ Grant for Activities Program (Project # 582-10-94307-FY14-06), TCEQ Supplemental Environmental Program (SEP Agreement no. 2009-009), and the Texas Air Research Center (TARC Grant #079LUB0096A) are gratefully acknowledged.

Special thanks are due to Ed Fortner and Scott Herndon of Aerodyne Research, Inc. (ARI) for providing numeric soot data for 2010 JZ flare campaign.

15AIChE
Annual Meeting, Salt Lake City, UT

Thank you!

424723 Quantitative Risk Assessment of Underground Propylene Pipeline: Deterioration and Corrosion Effects

Abstract:

Gases and chemical materials have been transferred by underground pipelines in petrochemical complexes. In Ulsan petrochemical complex, the largest one in Korea, the underground pipelines are laid since 1962, thus the oldest ones are superannuated more than 50 years. Moreover, the pipes have been rusted by soil. These deterioration and corrosion increase the possibility of accidents. In Kaohsiung, Taiwan, 20 years old underground propylene pipeline exploded causing 32 deaths and 321 injuries, last year. To prevent these accidents, the effects of deterioration and corrosion in underground pipelines should be studied quantitatively.

In this study, quantitative risk of underground propylene pipeline is analyzed according to deterioration and corrosion. The actual Ulsan petrochemical complex's information, such as geometry, population and weather, are used to calculate the risk. First, indirect assessment, such as CIPS (Close Interval Potential Survey) and DCVG (Direct Current Voltage Survey), and direct assessment are carried out to find out actual state of the pipelines. Second, accident frequencies are computed reflecting direct and indirect assessment by referring to OGP (Oil and Gas Producer) data. Then, consequences of underground propylene release are simulated in Phast v6.7. Next, release cases of new pipeline and old one are compared. As a result, adequate replacement schedule of the pipeline to reduce the risk is suggested. This research would contribute to the risk reduction in the underground pipeline explosions.

424746 Assessment of the Integrated Health Effects of Airborne Industrial Emissions Based on the Chemical Mixture Methodology

Abstract:

It is critical for Environmental Health and Safety agencies to have a consistent and comprehensive method to assess the health effects and the environmental impact in the case of exposure to mixtures of several emitted chemicals. Such an incident could emerge after an industrial accident but also due to continuous, though regulated, releases from large industrial cities. Today, there are approaches to estimate the combined effects of exposure to a mixture of chemicals. Usually, these approaches focus on predicting the impact from non-routine chemical releases, i.e. accidental releases. However, even regulated routine releases could pose a significant threat to human health when one considers the integrated effects. Therefore, it is important to explore the available methods to assess the impact in the case of continuous airborne releases from industrial cities (air pollutants mixture; several industrial sources).

According to early discussions (Bliss, 1939), chemical interactions are defined as a) independent joint action, b) similar joint action and c) synergistic action. For obvious reasons, many toxicological studies have followed, which eventually forged the existing environmental and health legislation. Many of these efforts were initiated by organizations such as the US Environmental Protection Agency (EPA, US), Agency for Toxic Substances and Disease Registry (ATSDR) and the European Chemicals Agency (e.g. REACH and CLP regulations) (Monosson, 205). All efforts focused on the collection and set of the appropriate approaches to conduct such assessments following specific categorization methods (Rice, 2008). Generally, there are three methods to conduct the risk assessments for the multi emitted chemicals: a) whole mixture assessments (actual mixture of concern), b) similar mixture assessments and c) component by component assessments. The usage of each method strongly depends on the availability of the basic required information such as inventory data for the emitted pollutants, concentration of the released chemicals and the meteorological data the considered location.

One of the recommended and simpler approaches, to conduct health risk assessment of chemical mixtures, is the Hazard Index (HI), also adopted by US, EPA and in REACH. HI does not cover all aspects of exposure. Therefore, more advanced methods need to be selected. One is the Chemical Mixture Methodology (CMM), which is extensively used for emergency preparedness in the U.S. (Department of Energy, DOE), but it has not been designed for the assessment of health effects from routine and typical industrial emissions.

Present work discusses the details of CMM and explores the necessary tools and assumptions in order to apply the method on continuous atmospheric releases and multiple sources; for the health impact assessment of the chemical mixture. These tools include: a) models to predict the emission rates of the released pollutants, b) a dispersion model (AERMOD) to predict the concentrations of the pollutants at several receptor points and c) an in-house algorithm that deploys the various realizations of the CMM. Finally, we demonstrate the methodology on a fictitious industrial city of Middle East Area. The outcomes support the applicability of CMM and the proposed tools to account for continuous releases with two main findings. First, the likelihood of sever impacts – hazard – increases in the case of a mixture of pollutants. Second, the selection of base exposure limit is a critical factor, and it can drastically change the conclusions of the methodology, in other words, the assessment of risk.

Acknowledgements

This publication was made possible by a NPRP award [NPRP 7 - 674 - 2 - 252] from the Qatar National Research Fund (a member of The Qatar Foundation). The statements made herein are solely the responsibility of the authors. The HPC resources and services used in this work were provided by the IT Research Computing group in Texas A&M University at Qatar. IT Research Computing is funded by the Qatar Foundation for Education, Science and Community Development (http://www.qf.org.qa).

References

Backhaus, T., Blanck, H. and Faust, M., 2010, "Hazard and Risk Assessment of Chemical Mixtures under REACH," Sundbyberg.

Bliss, C.I., 1939, THE TOXICITY OF POISONS APPLIED JOINTLY1, Annals of Applied Biology, 26, 585-615, 10.1111/j.1744-7348.1939.tb06990.x.

Carpenter, D.O., Arcaro, K., Spink, D.C., 2002, Understanding the human health effects of chemical mixtures, Environmental Health Perspectives, 110, 25-42.

Monosson, E., 2005, Chemical Mixtures: Considering the Evolution of Toxicology and Chemical Assessment, Environmental Health Perspectives, 113, 383-390, 10.1289/ehp.6987.

Rice, G., Macdonell, M., Hertzberg, R.C., Teuschler, L., Picel, K., Butler, J., Chang, Y.-S., Hartmann, H., 2008, An approach for assessing human exposures to chemical mixtures in the environment, Toxicology and Applied Pharmacology, 233, 126-136, http://dx.doi.org/10.1016/j.taap.2008.05.004.

433570 PAH Exposure and LUNG Cancer Risk Assessment By Internal Dosimetry Metrics

Abstract:

This study deals with the assessment of cancer risk attributable to PAH exposure, largely connected to the increased use of biomass for space heating in Greece in the winter of 2012-2013. Towards this aim, several methodological elements have to be integrated Towards this aim, several methodological elements have to be integrated, including:

- PM measurements for the three main fractions (PM10. PM2.5 and PM1). This is essential, since smaller PM fractions have been associated to biomass burning . Measurements include two different sampling sites (one representative of the urban background levels and one for traffic), in order to differentiate the amount and the composition of PM attributed to contribution of different sources.

- PM chemical analysis, aiming at:

  - identifying their carcinogenic potency, through PAHs composition analysis.
  - identifying their origin, through levoglucosan and black carbon (BC) analysis. Levoglucosan is considered as the most specific biomass burning tracer, while BC is considered as product of internal combustion sources.

- Refining the exposure and associated risk assessment methodology. Exposed population was divided to eight age-groups, ranging from infants to adults (>21 years old). Different types of daily activities demand different levels of effort that correspond to different inhalation rates. For the estimation of human exposure to PAHs, weighted average daily inhalation rate (IR_i, m^3/day) was calculated for each age group. Calculations refer to the time-activity pattern of individuals for a period of one week. It is known that PM of different size interact differently with the Human Respiratory Tract (HRT). PM from biomass burning has specific characteristics, which have to be taken into account when estimating the associated risk. HRT particle deposition modeling was applied for the determination of PM deposition fraction (DF) to the three parts of the pulmonary system in order to estimate the internal dose of PAHs. Main mechanisms of PM deposition across HRT include diffusion, sedimentation and impaction. Secondary mechanisms involve interception and electrostatic deposition. Different HRT regions involve different deposition mechanisms, with regard to different PM size as follows: 1. Naso-pharyngeal region (or upper respiratory tract – URT): impaction, sedimentation, electrostatic deposition. 2. Tracheo-bronchial (TB) region: impaction, sedimentation, diffusion. 3. Pulmonary/alveolar (P) region: sedimentation, diffusion. PM deposition is affected by the particle properties (concentration and size distribution), air flow parameters (lung capacity and breathing frequency) and HRT physiology (structure and morphology). HRT deposition was calculated using the multiple path particle deposition (MPPD) v.2.1 model. Calculation of the overall toxicity of the mixture of PAHs is done using toxic equivalency factors (TEFs). The lung cancer risk attributed to PAHs depends on the amount of TEQ deposited across the middle (TB) and lower (P) HRT regions, calculated as the sum of the products of the different size fractioned PM mass deposited across the different HRT regions, multiplied by the Toxic Equivalent Concentration (TEQ) estimated for the specific size fraction. Genotoxic effects of PAHs were estimated by calculating the amount of TEQ deposited across the middle (TB) and lower (P) HRT regions to a slope factor of 3.8×10^{-6} (kg×day)/ng B[a]P, calculated from the respective Inhalation Unit Risk for B[a]P ($IUR_{B[a]P}$), of 1.1×10^{-3} $m^3/\mu g$.

The latter allows us to estimate for a PM fraction of given size the actual amount of PAHs coming in contact with the lower respiratory tract. Using this refined exposure assessment approach, differences in physiology among different age groups/susceptible populations are taken into account in the corresponding cancer risk estimates.

Results showed that PM concentrations were higher in Greece during the cold months of the year, mainly due to biomass use for space heating. PAH and levoglucosan levels were highly correlated, indicating that governed the genotoxic effect of exposure to PMs.

Different size-distributed particle fractions are deposited across the different HRT regions, as a result of the different deposition processes occurring in the different regions of the HRT, which are related to the physiology/morphology of the respiratory system and the PM size distribution. This has been verified by a variety of lung deposition computational and experimental efforts. HRT deposition results indicate that the lower respiratory tract of infants and children (up to 14 years old) can retain up to 74% higher mass fraction of $PM_{1.0}$ particles than that of adults. The maximum difference in the thoracic deposition between adults and children between 3 and 8 years old reaches 68 and 230% for the $PM_{2.5}$ and PM_{10} fractions respectively. Thus, the $PM_{2.5}$ and $PM_{1.0}$ fractions rather than PM_{10}contribute to a greater extent to the absorption of PAHs by the respiratory tract in younger individuals than in adults. Although particle deposition in infants and children has been poorly studied and a high variability between individuals has been observed, key factors that influence the thoracic deposition of particles in children are lower ventilation rates (l/min), which favor deposition by sedimentation, a process that governs the deposition of larger inhalable particle diameters resulting in increased overall particle deposition profiles, as well as lower functional residual capacity FRC volumes.

Individuals who reside near the urban background site experience a higher risk (above 10^{-6}). Individual lifetime risk is higher for children, due to the higher bodyweight normalised dose of PMs and the amount of PAHs deposited at lower HRT regions. A higher fraction of smaller particles is deposited across the children respiratory tract, resulting in higher PAH burden.

The estimated lung cancer risk was non-negligible for residents close to the urban background monitoring site. Higher risk was estimated for infants and children, due to the higher bodyweight normalized dose and the human respiratory tract (HRT) physiology. HRT structure and physiology in youngsters favor deposition of particles that are smaller and more toxic per unit mass. In all cases, the estimated risk (2.8×10^{-6} and 7.6×10^{-7} for younger infants and adults at the urban background site and 9.5×10^{-7} to 2.3×10^{-7} for the traffic site, respectively was lower to the one estimated by the conventional methodology (4.7×10^{-6} and $1.6\times10^{-6}$for the urban background and the traffic site respectively) that is based on Inhalation Unit Risk; the latter assumes that all PAHs adsorbed to particles are taken up by humans.

This study provides new information on the toxicity of PM emitted from biomass burning based on a refined exposure and risk assessment methodological framework which takes into account the mechanisms of internal exposure to particulate matter. On this basis we could estimate health risk taking into account not only the spatial distribution of airborne particles, but also variance observed among different age groups. Our results suggest that the integrated mechanistic approach developed herein describes in more detail the interaction between toxicants adsorbed onto airborne particles and health effects on the urban population. This is clearly illustrated by comparing the relative risk between the people living in the two different sampling sites of our study. The non-mechanistic methodology that takes into account only *TEQ* and *IUR*reflects only differences in ambient air PAHs/TEQ levels. The mechanistic methodology proposed herein, incorporates the effect of complex environmental, physicochemical and physiological processes. These include:

(a) the PM size distribution emitted from biomass burning;

(b) the PAH adsorption on the different particle size fractions;

(c) the deposition distribution of different PM size fractions across the human respiratory tract.

The PM size distribution affects the way these are distributed across the different HRT regions. In turn, the deposition profile of PM across the HRT varies with age of the exposed individuals due to age- dependent differences in respiratory physiology. Thus, risk assessment can take into account that a higher fraction of small diameter particles (also found to be more toxic per unit of particles mass) are deposited in the middle and lower HRT. By incorporating all these parameters, the risks associated to the increased PM and PAHs due to biomass burning are further differentiated by source and age of the exposed population. Using the *IUR* methodology, cancer risk estimates for the population living close to the two sampling sites differ by a factor of three. This reflects only the difference of PAH levels found on PM10. With the methodology proposed in this work, the estimated cancer risk at the two sites differs by a factor of five to seven depending on the age group. Thus, the *IUR* method would result in significant underestimation of the actual cancer risk related to the particulate matter from biomass burning.

A Flare and Overpressure Management System

Part I: Methods, Metrics, KPIs and Software Solutions

Enhancing PSM by way of Focused, Metric-Driven Management Systems…a compilation of best practices made better

Mike Marshall, PE
Oil Industry Consultant
mtmarshall.llc@gmail.com
www.mtmarshall.net

This paper identifies the attributes and benefits of a data and metrics-driven management system focused on the process safety design integrity, reliability, and control of process plant flares and pressure relief systems. This management system process focuses on the four key ***business drivers of risk, regulatory, operations, and profits***, and involves several distinct ***business methods involving people, processes and tools/technology***. At the center of the management system is the unique design and implementation of ***metrics and KPIs*** created from data lifted and aggregated from an enterprise informational management platform.

1. About the Current Regulatory Climate

The highly publicized incidents at BP Texas City in 2005, Tesoro Anacortes in 2010, and Chevron Richmond in 2012 all happened not due to a failure of equipment, instrumentation, facility siting, operator, procedure, communication, supervision, or training, but rather a failure of all those things together, i.e., ***a management system failure***.

The BP, Tesoro and Chevron incidents are now driving the reexamination of the PSM rule by the US regulatory community. The US Chemical Safety Board (CSB) has taken notice that ***US Oil & Gas industry losses are highest among any industrial sector, and that the US Refining industry accident rate is 3 to 4 times*** higher than in Europe.

The PSM rule and its allegedly "less rigorous regulatory framework" are quickly falling out of favor with regulators. As such, the attributes of the "Safety Case" and ALARP regulatory regime currently in use throughout the United Kingdom, Australia and Norway are now being advocated by the CSB. And even more notable is California's proposed regulation for ***inherently safer design (ISD)***, an initiative which was just endorsed by CSB chairman Dr. Moure-Eraso, with him suggesting that other states should do the same.

ISD has been hotly debated for years, and would require that risk be reduced to the greatest extent possible with the selection and implementation of changes in chemistry and/or a change to process variables, e.g., reduction in pressure, temperature, flows, etc. Unmistakably, this would be taking the petrochemical industry and its PSM process safety approach ***from performance-based to prescriptive***.

Yet, before opting to prescriptively rewrite the PSM rule, I suggest that there is a performance-based option which is more sensible, productive and achievable in the short term, and that is a ***focused metrics-driven management system approach***. Such an approach also embodies the core principles of the PSM rule and is consistent with the findings and recommendations of the 2007 Baker Panel Report.

It would seem that the Baker Report is prompting us to revisit the PSM rule for intent and direction as well as ***the proper administration of the PSM rule***, i.e., the effective application of a management systems approach to continuously improving our process safety environment and culture. After all, the authors of PSM took great pains to make it a performance-based standard for a reason (i.e., prescriptive is inherently inferior), so let's not give-up on that now.

2. Revisiting PSM, Management Systems and Continuous Improvement

With promulgation of the PSM Standard 29 CFR 1910.119, OSHA mandated that a management system comprised of several well-defined elements be established *"for preventing or minimizing the consequences of catastrophic releases of toxic, reactive, flammable, or explosive chemicals."* The Process Safety Information (PSI) element of the PSM rule states *"the employer shall document that equipment complies with recognized and generally accepted good engineering practices (RAGAGEP),"* with specific reference given to *"relief system design and design basis."*

Although OSHA does not explicitly use the term *continuously improving* in their regulatory standards, they use equivalent terms such as *accurate, complete, clear* and *on-going*. For example, in the Appendix C compliance guidelines of 29 CFR 1910.119, OSHA uses the term *"complete and accurate"* in lieu of *"continually improving."* Likewise, for the Mechanical Integrity element of 29 CFR 1910.119, OSHA uses the term *"on-going"* to describe the expectation to continually improve.

Recent incidents and enforcement actions demonstrate OSHA's expectation for operating plants to maintain a continually improving PSM system. In 2007, OSHA initiated a special enforcement initiative (OSHA's National Emphasis Program [NEP]) specific to refineries and chemical plants. Of the citations issued, many involved missing, inaccurate and incomplete process safety information as well as outdated relief system studies.

Furthermore, flare and relief system design compliance is assessed annually in ever increasing detail for participants of OSHA's Voluntary Protection Program (VPP). ***Clearly, it is OSHA's expectation that operating plants have a continuously improving***

management systems process for ensuring the complete, accurate, clear and on-going integrity of design and operations of flares and relief systems.

And, besides process safety, there is the environmental enforcement aspect to flare operation and systems management which continues to make the news…

EPA: NSPS Subpart Ja (consent decrees and penalties)

- Equistar Chemicals – $1,900,000 (2007)
- INEOS (Lanxess) – $3,100,000 (2009)
- Marathon Petroleum – $460,000 (2012)
- BP Whiting Refinery – $8,000,000 (2012)
- Shell Deer Park - $115,000,000 (2013)

Besides, in the interest of social responsibility, it is just good business to develop a management system which not only enhances safety and environmental protection, but augments asset protection as well. Clearly, safety and environmental stewardship are of paramount importance, but asset protection, business continuity and public image also have vital significance in any business environment.

Of course, knowing what data to capture and display is essential to proper metrics development and analysis, and the ensuing derivation of key performance indicators (KPIs). So, what follows here is the ***who, what, when, where and how*** of doing just that with a focused, metrics-driven ***flare and overpressure management system, i.e., FOMS.***

3. A Management System Just for Flares and Relief Systems (FOMS)

It goes without saying that flare and relief system design and operation have come under intense scrutiny by OSHA. And now with the EPA getting into the act, it would seem as though regulators are now looking for a new '3% IPD' (inlet pressure drop) style soft spot, and found it in flare system operation and management.

The vast majority of time, gas flaring is associated with plant upset, misoperation or imbalance, and as such is unplanned and subject to regulatory penalty. EPA is now aggressively mandating and enforcing flare management plans (FMPs) and flare gas recovery systems just like OSHA enforced relief system pressure relief analyses (PRAs) and the 3% IPD rule.

So, it is clear that this double barrel surge by way of OSHA and EPA has arrived and is progressing rapidly forward. It may seem as though industry is under assault, but the ultimate truth is that ***process safety just makes good business sense***.

When considering a performance improvement program in this highly regulated process safety environment, ***four key business drivers should first be considered, i.e., risk, regulatory, operations, and profits***. Then, building a focused flare and relief systems management process around those four drivers involves a unique ***management system structure of people, processes and tools/technology***.

Central to the growth and continuous improvement of those three elements will be the proper design and implementation of metrics and KPIs. ***It will be the institutionalization of KPIs and the subsequent reporting and action planning process which drives the continuity and sustainability of the plan-do-check-act rudiments of this management system approach.***

4. Sustainability thru Key Performance Indicators (KPIs)

The PSM standard is exceptional in its vision, design and implementation, but it could have been made better by the inclusion of metrics and KPIs. As is often said, ***"if it can't be measured, it can't be managed,"*** and is likely a reason why so many process safety management system programs have failed to grow and measure up to industry best practices as well as OSHA expectations.

KPIs are the critical life's blood of a properly designed management system in that they institutionalize processes and drive accountability, which in turn provides for continuity and sustainability. An effective KPI system and data mining process takes into consideration business drivers, success factors, targets, improvement actions, and performance measures. But knowing which metrics should be funneled into KPIs is the challenge.

Again, it would now seem that ***API 754 was written only to gauge the 'high level' effectiveness of PSM programs***. And yet, opportunity still remains for developing more in the way of focused metrics which further drive performance improvement in areas like flare and relief system design and operation, among others. And correspondingly, the CSB has characterized the shortcomings of API 754 as follows:

1. Tier 1 and 2 numbers are lagging indicators and thus of limited usefulness as performance indicators
2. Statistical power of likely small numbers of Tier 1 and 2 events is insufficient to detect effect
3. Tier 3 and 4 events are leading indicators which are reflective of process failures, and yet are not publicly reported and utilized for industry trend analysis and benchmarking comparisons
4. Employee participation was insufficient in the development process and thereby lacking in a broad-based consensus

What I am suggesting is that we as an industry can be even more critical and innovative by utilizing historical operations, reliability and maintenance data in analytical tools and performance metrics to create a competitive environment for improving plant reliability and profitability. I stress the word 'competitive' in that this plan-do-check-act process will drive itself and grow by fostering a healthy and productive incentive among stakeholders for continuous improvement in reliability, profitability, and most importantly, process safety.

With KPIs, the idea is to tap into the data rich potential of an enterprise information management (EIM) system. And from this data and informational structure is extracted the 20% of data that 80% of operators, engineers, managers and execs want to see, with the challenge being identifying that 20% of key information. And beyond that, further consideration is necessary for the more refined development of KPIs which then provide the need-to-know requirements of stakeholders at a ***'dashboard' level of awareness.***

What so many KPIs fail to do is drill down deeply enough to facilitate the identification of basic and root causal factors associated with problem solving for optimal performance. And ***there can be too many of these focused metrics***, with the pitfalls being much the same as usability problems associated with multiple alarms annunciating during a process unit upset, commonly referred to as alarm flood. And just as with too many alarms, poorly designed alarms and improperly set alarm points, ***metrics flood and confusion*** can set in and negatively impact the problem solving process.

Proper development, implementation and management of metrics and KPIs should involve many of the same concepts utilized in alarm rationalization and management, and is really more of an art form than many realize. It requires critical thinking and strategic design aptitude which draws on frontline-to-exec level appreciation for what good looks like. And ***what good looks like*** is what the FOMS developers had in mind for a flare and relief systems focused management process.

Too often, the process of data gathering and metrics reporting is more about presentation than substance, and lacks real problem solving and process optimization potential. The metrics and KPIs of FOMS are specifically designed for problem solving performance improvement issues at the basic and root cause levels, and ***built around the business drivers of risk, regulatory, operations, and profits.***

5. The Profit Motive: Analyzing and Trending Performance

There is no doubt that profits are the motive behind all of business, and that includes the business of refining process safety. There are two primary objectives of most refining business strategies, and that is reducing costs and improving equipment reliability and asset performance. And as such, ***process safety is a cost of doing business*** which involves a risk versus reward calculation, i.e., consequence versus likelihood of a process safety event.

Two key indicators used to evaluate manufacturing cost effectiveness are mechanical availability and maintenance costs as a percent of replacement asset value (RAV). It is widely accepted by manufacturing companies that world class manufacturing performance means operating at greater than 97% mechanical availability and spending less than 2 percent on maintenance as a percent of RAV.

Therefore, tools must be used to analyze and trend performance relative to those two measures, and employ various 'deep dive' methods and indicators which drive toward root causes of inadequate performance. To that end, I have developed a proprietary refining incident and loss database as well as process optimization methodology (via RCFA) which quantifies the economic impact ($'s lost profit opportunity LPO) of equipment anomalies, LOPC incidents and upset/malfunction operation conditions.

Such a RCFA approach is key to analyzing and trending cost minimization, driving process optimization and maximizing process safety performance in the refining industry.

6. Management System Design and Implementation

As previously mentioned, PSM was conceived out of a management system mentality of a plan-do-check-act cycle with continuous improvement at its core. The focused, metric-driven management system of FOMS follows this same model and function. And in application, it begins with a three phase development process: ***where are we now, where do we want to go,*** and ***how are we going to get there***.

Phase I. Where are we now?

- Identify and engage process owners and stakeholders
  - A changing PSM and PRA landscape
  - BP, Tesoro, Chevron incidents driving reexamination of PSM rule
  - US refining accidents 3 to 4 times Europe
  - ***Safety Case and inherently safer design/technology (ISD)*** gaining favor with regulators
- Compile available documents and info
- Flowchart current processes, tasks and procedures
  - PRA methods and processes now mature
  - PRAs giving way to enhanced auditing, mini-PRA tune-ups and MOC processes
  - Is more needed to insure PRA integrity?
  - Intense regulatory scrutiny remains... ***risk, regulatory, operations*** and ***profit*** drivers
- Identify current tools and technology
  - PRA science and technology still evolving
  - Little in way of PRA specific management systems tools/IT
  - Lots of IT structure in need of management system content and integration
- Understand strengths, weaknesses, opportunities and threats in existing processes (SWOT analysis)

Phase II. Where do we want to go?

- Engage process owners and stakeholders for vision, objectives and value drivers
  - Business focus (but not safety second)
  - Drive subject matter savvy to front-line
  - Team environment, but competitive
  - Systems thinking and problem solving
  - Communities of practice and pride
  - Knowledge managers, not tribal
  - Bottom-up, top-down...invert pyramid with 'closest to the work' mentality
  - Measurements, accountability and rewards
- Baseline processes and perform gap analysis
- Evaluate gaps and tradeoffs (costs)
- Redesign processes and functionalities
  - Think like an operator, manager, regulator
  - Metrics and reporting...KPIs
  - Ongoing gap analyses...data centric
  - Expert systems to automate
  - Integrate with existing...customizable
  - Process optimization and profits
  - Better manage and control change
  - Enable regulatory compliance…Safety Case?
  - Quality processes...TQM, SPC, Six Sigma
  - Cross-org integration and collaboration
  - Focus on operations workforce
  - Standardization and consistency
- Specify tool and technology needs
  - Workflows
  - Protocols and practices
  - Portals and links to data and systems
  - Data repositories
  - Search engines and links
  - Dashboards, scorecards, forums
  - Executive dashboards
  - Document management
  - Training and more training (CBTs)
  - Enterprise discoverability and sharing
  - Design to drive sustainability
- Develop project plan and prioritize

Phase III. How are we going to get there?

- Identify needs and objectives
  - What good (or "where we want to go") looks like and how we plan a path forward

  - Critical focus on management systems design and implementation: people, processes, tools/technology (including software solutions)
  - Strategic focus on gaps, soft spots and critical systems within operations, maintenance and engineering organizations plus corporate
  - Content…the 20% of data 80% of stakeholders want to see (strategic and customizable KPIs, data maps, scorecards, dashboards, reports, data portals, alerts, analyses, trends)
  - ***Design to involve no new manpower, only new processes and tools***
- Develop strategic purpose
  - Maintain a business perspective on everything, including process safety
  - Tightly integrate strategy and tactics with business processes to be self-sustaining
  - Ensure organization and systems are designed to enable execution of business processes
  - Showcase new philosophy to inspire folks at all levels...from both bottom-up and top-down
  - Design for employee involvement for buy in at all levels...again, make it competitive
  - ***Integrate with existing assets, programs and systems***
  - Get KPIs in hands of those closest to the work, i.e., those most able to affect change
  - Connect enterprise performance measurement with budgets, reviews, and bonuses
  - Make everyone an ambassador, including regulators...especially regulators
  - Design metrics to be what operators, managers and executives want to see, i.e., KPIs
  - Make adaptable to and compatible with outsourcing applications of IT and engineering services
- Establish team, leadership and governance…culture is very important
  - Understanding and leveraging nuances of culture
  - Plant environment (operations, maintenance, engineering, corporate)
  - Cost/safety prioritization…RAGAGEP benchmarking
  - Address internal competitiveness and silos
  - Integrate with RBI and PSM offerings, leveraging IT and maximizing synergies
  - Get regulators on-board…local and state OSHA, EPA
  - Communicate to and involve all at all levels, and invert the hierarchical pyramid
- Perform root cause analysis
- Design metrics, KPIs, reports… automation tools
  - Design to drive sustainability (training, auditing, certification, profits)
  - Integrate with existing IT structure and software…synergies
  - Provide for enterprise discoverability and sharing
  - Leverage EIM platform, and integrate with:

        - Pressure relief analyses (PRAs)
        - Asset integrity management systems
        - Continuous emissions monitoring systems (CEMS)
        - PSM suite of software
        - Digital control systems
        - Process instrumentation
    - But, IT & software prowess need content in cohesive processes…FOMS
    - New/improved software solutions, business methods and internet innovations
- Initiate training programs
- Implement transition plan, pilot, then rollout
    - Compatible with 3rd party applications, software and systems
    - Leverage synergies/overlaps with PSM, equipment inspection and reliability programs (RBI API 580/581, especially damage mechanisms [API 572] and LOPC)
    - Flare/relief system specific programs for mechanical integrity, PHA, MOC, incident investigation, procedures, PSI, PSSR, and other PSM elements

Phase IV. **How do we improve, grow and keep going?**

- Implement and validate redesigned process
- Initiate ongoing metrics and management system
- Monitor, evaluate and report on new processes
- Review targets and performance
- Audit and adjust for continuity, sustainability and growth

7. Conclusion

In that a flare and relief system and associated safety systems are the last line of defense in overpressure protection, what better application is there for just such a strategic metrics-driven management systems initiative like FOMS. I will follow-up this paper with another entitled **A Flare and Overpressure Management System – Part II: Drilling Down into FOMS Enterprise Flare Manager Modules, Metrics and KPIs** which will further delineate the design and operation of the FOMS modules as well as discuss the application of associated metrics and KPIs.

8. A Close Second…

A close second to FOMS, however, would be a focused, metrics-driven management system approach addressing mechanisms contributing to the loss of primary containment (LOPC). The same people, processes and tools/technology (software and EIM) structure and methods employed in a flare and overpressure management system could be easily adapted for a LOPC focused initiative. And, this strategic initiative would involve the same business drivers of risk, regulatory, operations and profits.

What I am suggesting is that we as an industry can also be more critical and innovative in responding to LOPC incidents, data and metrics with enhancements to mechanical integrity proficiencies relative to inspection, maintenance, design and overall systems management. Historical operations, reliability and maintenance data can be better utilized and managed with analytical tools and performance metrics to determine needs, risk exposure, provide direction, and address opportunistic reliability issues. This would certainly include a more critical focus on inherently challenging API 754 process safety event (PSE) LOPC metrics relative to damage mechanisms, operating envelopes, and consequences of deviation, procedures, design and training.

LOPC is preventable, and equipment reliability relative to process safety is by far the leading risk opportunity and ongoing business concern facing the Oil & Gas industry today. A LOPC focused metrics-driven management system would provide a more robust program for facilitating comprehensive process design and engineering relative to reliability improvement, i.e., ***'safety by design.'***

I am currently writing a paper on the topic of LOPC and safety by design. The paper will also feature my proprietary refining incident and loss database as well as my process optimization methodology (via RCFA) which quantifies the economic impact ($'s lost profit opportunity LPO) of equipment anomalies, LOPC incidents and upset/malfunction operating conditions.

430939 A General Model for Estimating Air Emissions in Petroleum Refineries Using Emission Factors Technique

Abstract:

A General Model for Estimating Air Emissions in Petroleum Refineries Using Emission Factors Technique

Ba-Shammakh, M. S

Chemical Engineering Department, King Fahd University of Petroleum & Minerals

P.O. Box 5050, 31261, Dhahran, Saudi Arabia

ABSTRACT

Estimating the quantities of certain substances discharged into the atmosphere by several industrial activities is a necessary and fundamental stage in any environmental protection policy to tackle today's problems such as acid rain, the degradation of air quality, global warming and climate change and the depletion of the ozone layer. Any environmental protection policy requires comprehensive information on the pollution profiles of particular activities. In most countries and for petroleum refineries, particularly, such comprehensive information base does not exist. This paper proposed a general model for filling part of the information gap with respect to air emissions estimation from petroleum refineries. It presents a pollution-estimation model that draws from published sources of information and allows analysts to obtain order-of-magnitude estimates of the air pollution generated by refineries. As an illustration, the model is applied to a case study.

AUTHOR INDEX